B. F Underwood

Headache and its Materia Medica

B. F Underwood

Headache and its Materia Medica

ISBN/EAN: 9783743330689

Manufactured in Europe, USA, Canada, Australia, Japa

Cover: Foto ©berggeist007 / pixelio.de

Manufactured and distributed by brebook publishing software
(www.brebook.com)

B. F Underwood

Headache and its Materia Medica

HEADACHE

AND ITS

MATERIA MEDICA.

BY

B. F. UNDERWOOD, M.D.

———

NEW YORK
A. L. CHATTERTON & CO.
1889.

PREFACE.

In preparing the following pages for publication, it has been the endeavor of the writer to collate the accumulated experience of the Homœopathic School of Medicine in such a shape as to place it readily at the command of the practitioner in the treatment of one of the most prevalent and painful maladies against which he has to contend. The sufferer from headache, too often believing that there is no balm in Gilead for his pain, bears in sullen submission the agony which is apparently inevitable, instead of seeking the relief which medical art should afford. It is probably true that in no other disease is careful discrimination and judgment more necessary in the selection of the remedy than in headache, if the best results are to be obtained; it is equally true that the similimum being found, no other disease is more amenable to treatment. If this work shall assist the physician, who has a doubtful or obscure case to treat, to find more readily and easily the similimum, its object will have been accomplished. How well it may serve this purpose must be left to the judgment of the medical profession.

In its preparation all available sources of information have been freely drawn upon, and the proper credit given throughout the work. If in any case it fails to appear, the omission has not been intentional. Among the many to whom credit is due, mention may here be appropriately made to, Day, Strumpel, Hale, Hering, Lilienthal, Hughes, Farrington, Hempel, Raue, King, and to current medical literature.

575 Classon Ave., *Brooklyn.*
December, 1888.

HEADACHE.

TO the habitual sufferer from headache, there is no other pain which at all compares with it ; the premonitory symptoms of a coming headache are replete with terror, and the suffering while it lasts is so severe that rather than endure its long continuance, death, and a cessation of pain, seems preferable. It may be doubted, if the discomfort and suffering from headache, although it be but one of the minor ills of life, does not in the aggregate exceed that from any other disease.

While in many, and perhaps in most cases, it is but a symptom, rather than a disease, yet to a certain extent headache constitutes of itself a disease, and is the principal, if not the only symptom of which complaint is made. The pathology of the cerebral lesions involved in headache is obscure ; the diagnosis difficult ; for hidden beneath its osseous encasement the morbid changes in the tissues of the brain or its functional derangement, except as revealed by the subjective symptoms of the patient, are as a sealed book. We can, however, recognize as underlying it, and as a prime factor in

its causation, a peculiar susceptibility of the nervous centers, which, under the stimulus of various
disturbing influences, results in headache.

According to the exciting cause from which they
spring, headaches may be classified in various divisions, of which the following are the best defined :
1, anæmic ; 2, hyperæmic ; 3, nervous ; 4, reflex ;
5, rheumatic ; 6, toxic ; 7, catarrhal.

ANÆMIC HEADACHE.—The first of these, the
headache of anæmia, is properly symptomatic. It
is produced whenever a sufficient supply of the red
corpuscles of the blood are not passing through
the capillaries of the brain. This condition may be
due to an alteration in the quality of that fluid as
in anæmia or leucocythæmia, where the number of
corpuscles is reduced below the normal proportion ;
or while the blood may be normally constituted,
the deficiency may be due to a weak action of the
heart. The diagnostic symptoms of this variety of
headache are such as would naturally arise from
deficient cerebral circulation. There are spells of
depression and lowness of spirits ; with fearfulness
and timidity. Dread of the future, of business or
of other troubles, which are unfounded. The pain
is usually of a dull, gnawing character, affecting the
vertex, which is hot, and burning to the touch. It
may also affect the forehead and temples and more
rarely the occiput. It is accompanied with pallor
of the face, and palpitation of the heart ; and often

associated with dyspepsia and cardialgia ; there is also coldness of the extremities, and in females there may be dysmenorrhœa. In some cases sleeplessness and in others drowsiness predominates. The organs of the special senses are disturbed, especially when loss of vital fluids has taken place, with ringing and buzzing noises in the ears, flashes of light before the eyes, and more rarely, illusions of smell. There may be also twitching of the muscles, dizziness and fainting. The digestion is often disturbed with nausea, furring of the tongue, flatulency and constipation. The eyes are sunken, the pupils dilated or in some cases contracted. The pulse is usually weak, labored and slow, or small, weak and rapid. In addition to the indicated remedy in this form of headache very much of the success in treatment will depend upon the diet and the surroundings. Change of scene and occupation, the securing of rest and quiet, may be absolutely necessary to effect a cure. If there are any drains upon the vital strength these will necessarily demand attention. A nutritious and restorative diet is also an important adjunct.

HYPERÆMIC HEADACHE.—This variety of headache, while the result of a directly opposite condition from that of anæmia, may be due to similar causes, such as prolonged thought and mental labor. Under the stress of severe mental exertion there is an increase in the amount of the arterial blood sent to the brain, the cerebral vessels are dilated and this condition remains when the demand for the increased supply has ceased. The blood-vessels are distended and there is a corresponding pressure upon the substance of the brain. As the hyperæmia is active or passive, arterial or venous, there will be a corresponding difference in the symptoms. These will vary somewhat, at times sensation being most affected, at others the mental faculties. In the latter case the thoughts are rapid, disconnected and changeable, alternating from grave to gay, and at times becoming maniacal. The ideas are confused, false and distorted. The entire head is usually affected with sensation of pressure and throbbing. There are illusions of the special senses from the irritation of the nervous centers, with extreme sensitiveness to noise of every kind ; light is unpleasant and aggravates the suffering. The eyes and face are suffused and there is strong pulsation in the carotids. The face is sometimes intensely red with heat in the brow and vertex. In some cases there is intense suffering, hyperæsthesia and violent pains.

When the headache is the result of severe mental application, anxiety or over-indulgence in rich food, or stimulants, a change in the habits is the first steps toward a cure. Rest and moderation in living should be insisted upon and attention paid to the diet. A light diet, fish, white meat, fruits and vegetables should be substituted in the place of the more nitrogenous foods and all wines and liquors should be avoided.

NERVOUS HEADACHE.—This form of headache is due to a morbid change of the nervous centers, a peculiar functional disorder which is paroxysmal and not continuous, and which is often inherited. To the paroxysms which are similar to those which result in neuralgia, epilepsy, etc., Dr. Edward Liveing has given the name of "Nerve Storms." "This consists in a tendency on the part of the nervous centres to the irregular accumulation and discharge of nervous force, to disrupted and unco-ordinated action, in fact ; and the concentration of this tendency in particular localities, or about particular foci, will mainly determine the neurosis in question." [Quoted by Day.] In case the general health has been reduced the attacks become more frequent and severe, the controlling influence which holds this tendency to morbid irregularity of action in check is weakened and the abnormal disposition has full sway. Under such circumstances the slightest irregularity of living will suffice to bring on an attack, fatigue, worry, excitement or indulgence at the table often being a sufficient provocative. The headache is ushered in by various premonitory symptoms which may precede the attack by a period varying from a few hours to several days, during which time there is a sense of general uneasiness and discomfort, pressure in the head, vertigo, ringing in the ears, spots before the eyes, chilliness, malaise and yawning. The pain is usually intense and is located in the forehead and vertex, but may attack any portion of the head or the back of the neck. After continuing for some time it often settles in one temple or in one eye or in one side of the head. Associated with these headaches we often find disturbance of the digestion, nervous irritability and disquietude. The pulse is generally undisturbed even during the most severe paroxysms, although in some cases it is feeble and small. The pupils are usually contracted and the extremities cold with a sensation of heat in the head.

HEMICRANIA.—This is a severe form of nervous headache due to vaso-motor disturbance, or associated with vaso-motor symptoms. The pain is felt most in the anterior frontal region, it is generally of a continous character, not intermittent, and may increase to great intensity. Special painful points, such as are often present in neuralgia, are usually absent, but the skin over all of the affected part is generally hyperæsthesic. The malaise of the premonitory period continues, the sufferer has no appetite, there is often nausea and extreme sensitiveness to external impressions.

According to M. Hervez (*Practitioner*), "Hemicrania is an arterial neurosis which takes its origin in the great sympathetic nerve, and its seat is in the nervous filaments which accompany the arteries, whilst it manifests itself in the dilatation of these vessels, and in the compression of the brain and other organs it produces."

This form of headache may be divided according to the vaso-motor symptoms into two divisions—"*hemi-crania sympatheticus tonica or spastica* and *hemi-crania sympatheticus paralytica or angio paralytica.* In hemicrania spastica first described by Dr. Du Bois-Reymond, the forehead and ears on the affected side are pale, the temporal arteries contracted, the pupils often decidedly dilated, the secretion of saliva is increased ; in short, there are a whole row of symptoms present which all agree in pointing to condition of irritation in the sympathetic. In hemicrania paralytica first described by Mollendorf, also from observations on himself, the face is reddened on the affected side, it feels hot, the temporal arteries are dilated and pulsate strongly. There is sometimes unilateral sweating of the face, the pupil is contracted—all symptoms which can only depend on a paralysis of the sympathetic." (Strumpell). We, however, find that often these opposite conditions alternate or appear at the same time. The origin of the pain remains unexplained, it may arise in the disturbance of the circulation or from pressure of the distended vessels upon the nerve trunks.

NEURALGIC HEADACHE.—As a variety of nervous headache, neuralgic headache may be considered in this connection. This, like hemicrania, usually affects one side of the head and face, or locates in one more or less circumscribed spot, causing a sensation as though a nail were being driven into the brain. It frequently proceeds from some local cause of irritation, as a decayed tooth, or it may result from fatigue, excitement, worry, from gastric irritation, or from cold or exposure. In some cases the disease is intermittent in character, recurring at regular intervals, or at certain periods. In other cases the attacks occur irregularly. The true neuralgic headache seldom extends over the whole head, or produces sickness or vomiting unless the attack has lasted long and is very severe. "I should limit its strict definition to the intensity of the suffering, and its superficial seat ; to the paroxysmal character of the pain, and its extension in the course of the superior branch of the fifth nerve, and those filaments which supply the orbit, inner angle of the eye and forehead."—(Day.) The pain in some cases seems to extend through the head from front to back, or to pass through the eye or the temple and emerge at the back of the head, and in some instances through the neck or extremities. The pain being described as being similar to a hot rod or wire passing through the brain. Numbness and tingling in various parts of the body or limbs often accompanies the pains. Nausea and vomiting, when present, follow the pain, and are the result, not the cause, and do not relieve as in reflex headache. Although the retching may be severe, there is vomited only a little frothy mucus.

As neuralgic headaches are frequently the result of anæmic conditions, often arising from mental strain or overwork, attention to the environment of the patient is requisite, and rest and a nourishing diet are important adjuvants to treatment.

REFLEX HEADACHE.—Reflex or symptomatic headache is another variety of nervous headache, the exciting cause of which is an irritation of the nervous system often remote from the seat of the pain. It may be in the stomach, as in the so-called bilious headache, or in the uterus in the menstrual headache. It may be doubted if the uterine or gastric irritation would be sufficient to produce the headache if the abnormal sensitiveness of the nervous centres were absent, for uterine, gastric or other irritations may exist without any cephalic disturbance whatever. An explanation of this morbid susceptibility of the nervous system to the irritant in these cases is lacking. At certain times and under certain circumstances this susceptibility becomes more marked, so that headache will follow upon trifling derangement of the general system. A frequent source of this form of headache, and

one often overlooked, is eye-strain, due to a disorder of accommodation or an insufficiency of one of
the ocular muscles.   It is usually frontal, or in the
region of the eye.   The pain is increased by the
use of the eyes, and, in the beginning, follows upon
excessive use of the eyes, as after an evening spent
at the theater or exposure to bright light, although
later it often appears apparently spontaneously.   In
many cases there will be found characteristic symptoms, chronic irritation of the conjunctiva, with
intense redness and velvety appearance of the
mucous membrane and of the tarsus, which greatly
assists in the diagnosis.

DYSPEPTIC HEADACHE has its origin in imperfect
digestion and arises either in the stomach or in the
duodenum.   The pain comes on in the morning
after a hearty meal, or after drinking too much
wine.   The pain is of a continued shooting character, which may be diffused over the whole forehead and top of the head, or concentrated into
a small spot.   If diffused, the head feels hot
and burning, the face is flushed, with throbbing
of the temporal arteries.   The pain is aggravated
by motion or heat, which produces nausea, and the
attack finally culminates in a fit of vomiting, which
is followed by heat and a cessation of the pain.
The duration of the attack is variable ; in light cases
lasting only a few hours, while the severe form may
continue for several days, changing during its

course from a bilious to a nervous, or from a nervous to a bilious headache.

The following indications have been given as diagnostic of the seat of the irritation in reflex headache :

I. When the pain is located between the ears at the occiput, below the lambdoidal suture, the gastro-digestive apparatus, the automatic centers of life and the sexual organs will be the seat of disturbance.

II. When the pain is located in the region of the parietal bone, from the coronal to the lambdoidal suture, and from the squamous suture to the superior outline of the parietal eminence, the duodenum and small intestines will be the seat of disturbance.

III. When the pain is located in the forehead, from the coronal suture to the superciliary ridges below, and within the temporal ridges on either side, the large intestines will be the seat of disturbance.

IV. When the pain is located below the superciliary ridges, including the upper eyelids, to the external angular processes on either side, the nasal passages and buccal cavity will be the seat of disturbance.

V. When the pain is located in the temporal fossa, from the squamous suture to the zygoma below, and from the temporal ridge to the mastoid process, the brain and meninges will be the seat of disturbance.

VI. When the pain is located at the vertex, from the coronal suture and two inches posterior to it in the median line, and two inches on either side of that extent, in the female the uterus, and in the male the bladder, will be the seat of disturbance.

TOXIC HEADACHE.—Headache due to the absorption by the system of a poison is merely a symptom of a general disease to which treatment should be directed. The most common forms are those arising from alcoholic, narcotic, uræmic, or malarial poisoning, in which the symptoms will vary according to the poison absorbed. When due to alcoholic poisoning the headache is usually associated with gastric disturbance, and resembles a gastric headache.

In the malarial form the pain occurs in paroxysms, having more or less regular intervals, usually intense and developing at certain hours about one of the supra-orbital foramen. The pain continues for some hours, often of great intensity, and accompanied in some instances with fever, sweat, and other symptoms of malarial poisoning.

In nervous persons excessive indulgence in tea or coffee will frequently give rise to severe headaches, which may be often temporarily relieved by a cup of strong tea, but which continually recur, presenting the symptoms of a nervous headache, and which can only be permanently relieved by a total abstinence from these substances.

RHEUMATIC HEADACHE.—This form of headache, which is usually associated with the rheumatic diathesis, affects the fibrous tissue of the scalp, and the occipito-frontalis muscle. It is characterized by an aching pain, and heavy and continuous tenderness of the scalp and jaws. There may be aching pain in the teeth, which often become tender so that mastication is painful. The face is sometimes flushed, but the temperature of the scalp is unaltered and the temporal arteries do not throb, although the vessels of the head and face are distended. In some cases the pain is paroxysmal and hemicranial, especially affecting the forehead and vertex, from whence it radiates in various directions. It is increased toward evening and before a storm, growing less toward morning, so that the days are more comfortable than the nights.

Diet plays an important part in the treatment of this form of headache, and should be of a character adapted to the rheumatic tendency. The food should consist of fresh vegetables, fruits and milk in preference to animal food, and all liquors, except a little dry wine, should be avoided.

CATARRHAL HEADACHE.—This form of headache is due to a subacute inflammation of the mucous membrane of the nasal cavity and the frontal sinus, with swelling and dryness of the membrane.

The pain is located in the lower portion of the fore-head and in the root of the nose. It varies from a dull aching to intense shooting and burning pains, which radiate to the side and top of the head. It usually begins in the morning on awakening, and continues through the day with aggravation from motion. The brain is dull and heavy, with a disinclination for any mental or physical labor. If the pain is severe there may be some disturbance of the stomach, which, however, never amounts to vomiting. The mucous membrane of the nasal cavity is dry and burning. The attack may last from a few hours to several days, passing away with sneezing or profuse discharge from the nose.

BELLADONNA.

Belladonna, according to Dr. Hughes, is our chief remedy in headache. It is suited to the nervous or neuralgic form, as well as to the congestive. Heavy, drooping eyelids, and blindness or flashes of light before the eyes, point to it ; also flushed face, hot head, and sense of burning in the eyeballs. Secondary vomiting does not contra-indicate it ; but in true gastric headache it is of no use. The belladonna headache is always aggravated by light, noise, and movement, and also by lying down ; it is easiest in a quiet sitting posture. Its essential characters, indeed, are hyperæmia and hyperæsthesia.

Hempel speaks of the headache of belladonna as

frequently right-sided; it is usually very violent, and symptoms of congestion to the brain are well marked; vertigo, stupefaction, and partial loss of sight are frequent concomitants. The attacks are apt to come on or to get worse in the middle of the afternoon and to last until toward morning. Aggravations from noise, light, walking, jarring, motion of the eyes, touch, stooping, lying down, heat of bed. Ameliorations from quiet, sitting in the chair, dark room, strong pressure on the forehead.

Among the older German writers Black says, belladonna is adapted to headaches where there is present pressure, fullness, heat in the head, flashes of redness in the face, cold feet, roaring in the ears, partial deafness, and dilatation of pupils. It is adapted to persons predisposed to active congestions, to plethoric persons of a sanguine temperament, with great excitability of the uterine system.

Several forms of headache have been cured by belladonna. Sometimes, after taking cold, there was visible evidence of a rush of blood to the brain; a purple, swollen countenance, pulsating carotids, and a pain in the head as if it would burst. Again, the face was pale, cold, and there was a dull, unbearable pressure in the brain, as if it were cramped for room in the cranium. Vomiting quite often exists. [Heichelkeim.]

According to Tietzer, belladonna is indicated under the following conditions : The characteristic one-sided pain in the superciliary region, extending downward into the orbit and nasal bones, with increased lachrymation ; the pain is pressing in the forehead and eyes, and there is often a feeling as if the parts were being pressed apart. Every movement of the head and eyes increases the pain to a point of absolute anguish, especially when the eyes are exposed to bright light ; in the same manner every noise, and walking about aggravates the pain. The eye itself is often bloodshot, and there is much lachrymation. Stomach troubles, pressure, nausea, regurgitations, even vomiting appear sympathetically. Tietzer says that belladonna can only palliate, but not cure, cases where headache is secondary based upon primary gastric disturbances.

Belladonna cures when the head is sensitive externally, the blood-vessels of the head and hand congested, there is a swimming sensation in the head, roaring before the ears, darkness before the eyes, with the most violent one-sided headache, extending into the nose and eyes, with pressing, explosive, undulating, swaying sensations, worse from each motion, turning of the eyes, noises, jar, attempts to walk. With every step and with going upstairs it jerks and snaps in the forehead.

Pains coming on every afternoon, continuing until midnight, worse in the warm bed. Pains commencing very lightly, running into a stitch involving one-half of the head, so intense and so deep that consciousness is lost. [Hygea.]

In neuralgic headaches the pains come on suddenly, last indefinitely, and cease suddenly. Stabbing pains, as if with a knife, from one temple to the other. Great sensitiveness to cold air, can detect the moment a door or window is open in any other part of the house. Takes cold from every draught of air, especially when uncovering the head; headache from having the hair cut. Sleepiness, but cannot sleep; great restlessness with sudden starting. Ill humor, with paroxysms of rage with desire to cut or tear things, or to stab some one. Belladonna is suited to children, females, and young people of mild temper, blue eyes, blonde hair, delicate skin, and mild complexion. To women where the menses are early, with bright red blood. Pressure as though all the contents of the abdomen would issue through the genital organs, especially felt in the morning.

Case I. Mrs. B., aged 40, of stout build, had suffered for some time from a rheumatic pain in the foot for which she liberally painted the foot with iodine. This caused a cessation of the pain, but three or four days later she was taken with a violent pain in the head, which, although continuous, was markedly aggravated from about 2 P.M. until 6 A.M. Walking the floor all night, with frequent paroxysms of anger and desire to destroy something or to injure some one. Face pale, darting pains, with chilliness, desire to keep the head wrapped up, very sensitive to the least cold air, can feel the moment a door or window is opened downstairs, even with head closely enveloped in shawl. Belladonna tincture, 2m. in four ounces of water, teaspoonful every hour, gave decided relief, but pain was still felt on motion, laughing, etc. Belladonna 3 every two hours. The following day the headache had returned with more than former violence. Belladonna tincture, 4 m. in four ounces of water, a teaspoonful every two hours, gave entire relief, and neither the headache nor the pain in the foot have since returned.

Case II. Mrs. J., aged 40, slender, of a nervous, hypochondriacal temperament, was taken with a violent pain in the forehead extending through the head to the occiput. Worse from 3 P.M. to 6 A.M. Great sensitiveness to cold air, must have the head wrapped, a door or window opened in any part of the house is felt at once. Belladonna tincture, 2 m. in four ounces of water, gave prompt relief.

NUX VOMICA.

Nux vomica is, according to Dr. Jousset, the most frequently indicated remedy in migraine ; it is suitable in gouty and hemorrhoidal patients, who form four-fifths of all the cases. The migraine cured by nux vomica begins in the morning on walking and increases during the day; nausea and vomiting during the attack ; aggravated by intellectual labor, by motion and by rest ; extension of the pain to the occiput, where it frequently becomes more intense than elsewhere. It is best for men ; persons of sedentary habits who study much. Sensation as if the skull would split, worse during motion or stooping. Nux is a capital remedy in headaches caused by over-eating, abuse of coffee, spirits, excessive mental labor. The head feels as if it would split ; a sort of painful pressure with sticking pain. Catarrhal headache ; the brain feels heavy and aching, as if bruised. Rheumatic headache ; a tearing pain after eating, with sensation of heat in the cheeks, and a chilly feeling over the body, or only in the hands ; also with throbbing in the forehead ; or a crampy pain in the head, with soreness and sensitiveness of the scalp. Gastric or bilious headache, worse after eating, with nausea, sour vomiting, tearing and burning pains in the head and forehead, vomiting of bitter and sour phlegm.

According to Dr. Hughes, few medicines are more frequently indicated in headache than nux vomica. The headache of strong, plethoric adults, with congestion, giddiness, flushed face, and constipation, the pain increased by taking food or by mental exertion. But it is also [in higher dilutions] curative of such cephalalgia as clavus and migraine, where the constitution suits nux better than ignatia.

Tietzer says, regarding nux, the affection proceeds from the ganglionic system, and the hemicrania is sympathetic ; in hemorrhoidal constitutions, choleric temperament, persons accustomed to
drink freely, students with sedentary habits, etc.
The pain most frequently is a drawing pressing ; a
sensation as if a nail were driven into one side of
the head ; on the affected side the brain feels as if
crushed, bruised.   In women menses too soon and
too copious.   The headache comes on early, right
after a meal, after intellectual effort.

Knorre found nux vomica useful in periodical
headaches, commencing every morning after rising,
increasing until noon, then decreasing again, pressing and tearing, with and without complications,
such as pain in the hepatic region, nausea, inclination to vomit, bitter eructations and vomiting ;
hardness of abdomen ; choleric temperament.

Nux relieves headaches due to constipation, to
the use of coffee, also headaches as if caused by a
nail, or jerking stitching pain, with nausea and sour
vomiting ; when there is stitching and pressure on
one side, commencing early and growing constantly
worse, until he almost loses consciousness and raves
and tears ; any effort to think aggravates it so that
it feels as if the head would burst ; the brain itself
is sore, as if torn ; the face looks pale and haggard ;
the head feels heavy, with buzzing in it, with dizziness and trembling when walking ; worse from

motion, even of the eyes, in the open air, early in
the morning and after eating, or from stooping ;
if the head aches externally, cold aggravates it.
[Hering.]

According to Schroen, Hygea., young men and
girls who have masturbated often experience symptoms like these : Suddenly they perceive of objects
only the point upon which the vision is firmly fixed,
while all the rest becomes indistinct ; thus, of the
hand they see one finger ; of a chair one leg. Of
printed words they can read letter by letter, but
are not able to take in an entire word at a glance.
Appearance of bright light or fire from the canthi
toward the pupil, like a wheel of fire. Lachrymation. All this is followed by the appearance of an
intense, continually increasing headache, lasting
twenty-four hours, with vomiting. A few doses
of nux vomica will relieve and cure these paroxysms.

Nux, Lobethal says, is adapted to headaches
resulting from the use of spiritous drinks ; from
great mental efforts, connected with congestions.
especially in persons of sanguine, excitable temperament, and consisting of a distressing, indescrib-
able feeling of a dizziness mixed with stupor, particularly when there are complications such as pre-
cordial pressure, pain in the hypochondria, nausea,
hiccough, and constipation. Congestion to the
head, due to habitual obstruction or to disturbances

of the portal circulation, are more quickly relieved by nux than by any other remedy.

Thin, irritable persons, with vehement dispositions, or lively, sanguine temperaments. Patients of sedentary or intemperate habits, or those troubled with piles. The pains which come on by keeping one's-self confined in a room are relieved by a walk in the open air, and *vice versa*. Scalp sensitive to the touch or to the wind ; better from being warmly covered. Fetid sweat of one-half of the head and face, which is cold, with anxiety and dread of uncovering the head.; sweat relieves the pain. No desire to do any kind of work. Noise, talk, strong odors, and bright light are intolerable. Can not sleep after 3 A. M. ; ideas crowd upon him so as to keep him awake for hours.

Case.—A woman, aged twenty-seven years, of sanguine temperament, passionate, choleric, sensitive, suffered from sensation of wavy motion in the brain ; sensation in the head as if she had been on a spree ; pressive pain in the occiput early in the morning ; pressive, beating headache upon the slightest attempt to do a problem in arithmetic ; increased by coffee and wine ; after eating, drawing pain in the teeth and temples ; looseness of the teeth ; bitter taste in the mouth. Nux vomica 24 cured in six days. [Schreter.]

BRYONIA.

Bryonia is adapted to several varieties of headache, as gastric, rheumatic, catarrhal, and congestive, but particularly to gastric and catarrhal. The
pains are heavy and throbbing, or sharp, darting and
stitching ; and are often associated with nausea and
vomiting. The pains are worse from motion, and
better from quiet and in the open air. Wherever
there is pain there is usually soreness. Loss of
memory and inability to collect the thoughts. The
patient is very irritable, vexed, vehement. Rheumatic headache in cold, raw, wet seasons. Headache from ironing. The scalp is very tender to the
touch. Headache sets in on first waking in the
morning, or every day after dinner. Bryonia is
spoken of by Hughes as useful in hemicrania ; the
pain is generally on the right side, and is accompanied by retching and bilious vomiting.

According to Black, bryonia is the remedy when
the pain is on one side of the anterior portion of
the head, and extends to the neck, arms, face,
becoming more of a beating pain as it increases in
severity. In such cases it may act well in alternation with alumen.

The head symptoms of bryonia, together with its
physiological action in other parts of the system,
point to changes and disturbances in the sensorium
and in the functions of several cranial nerves of the

nature of a true hemicrania. It is in this respect
closely related to colocynth and nux vom. Under
colocynth, however, the motory nerves of the face
are not affected ; under bryonia those branches of
the superior facial nerve which supply the temporal
and maxillary region are not unfrequently affected.
[Oester, *Zeitschrift*.]

Knorre says that bryonia is adapted to headache
involving the frontal and temporal region, pressing,
crowding from within outward, as if the head would
bnrst, with sharp stitches so violent that the patient
screams, caused by cerebral congestion, and aggravated by stooping, motion of the head, coughing
and sneezing.

Bryonia relieves the burning or pressing pains,
with a feeling, when stooping, as if the contents
the cranium were being crowded out of the forehead, worse from walking, better from pressing
with the hands or bandage ; or a more external
"tearing," running into the face and temples, or
pressing, digging pain in small spots, especially in
rheumatic persons and those of an angry, irascible
disposition. [Hygea.]

Case.—A woman, æt. 30 years, very sensitive and
easily frightened ; menstruation formerly irregular,
often too copious, now normal ; has for a long time
suffered from violent headaches, particularly after
being subjected to depressing influences. The
headaches commence in the morning, when awak-

ing, increase during the day, and in the evening reach such a degree of intensity that she becomes actually wild with unbearable pain. Pain of a pressing-together character, involving the entire upper part of the cranium ; eyes dull, glassy, small, cannot be opened on account of the pain, can bear no noise or light. With it angry, quarrelsome, stubborn. In the afternoon, palpitation of the heart, dyspnœa, nausea, empty eructation. If she can get sleep during the night, she feels better on the following day ; but if fever occurs, she is for several days confined to the bed. Cured by bryonia 3. [Mueller.]

CIMICIFUGA.

Cimicifuga is indicated in nervous or hysterical women, especially at the menstrual period, and at the climateric ; rheumatic headache, affecting the muscles ; periodical or remittent nervous headaches and in hysterical or nervous headaches. It is also indicated in headaches resulting from loss of sleep ; from abuse of alcoholic drinks ; from mental strain and worry of mind ; from exposure of head to draughts of cold, damp air.

Nearly all the pains in the head extend to the eyeballs, and are attended by faintness, and " sinking " at the pit of the stomach. Pain over the eyes, and in the eyes, extending to the base of the brain, to the occiput. Severe pains over the right or left eye, extending to the eye and base of the brain, with dejection of spirits. Severe remitting headache of long standing, occurring every day at the same hour. Moving the head or turning the eyes causes a sensation as if the head was opening and shutting. Severe pain in the eyeballs, extending into the forehead, and increased by the slightest movement of the head and eyeballs. Intense throbbing pain, as if a ball were driven from the neck to the vertex, with every throb of the heart. Rush of blood to head, brain feels too large for cranium ; after suppressed uterine discharges, or suddenly ceasing pains.

The pains in the head seem to extend over and through the whole brain, producing a distinct sense of soreness in the occipital region, increased by motion. Congestion of the eyes during headache. Sensation as if a heavy, black cloud had settled all over her and enveloped her head, so that all was darkness and confusion, while at the same time it weighed like lead upon her heart. Sleepless from nervous irritation. Eructations, with nausea and vomiting, with headache. Sensation when going upstairs as if the top of the head would fly off. Pain in the back and along the spine are important concomitant symptoms.

Case.—Miss L., aged twenty-seven years, of dark complexion, nervo-bilious temperament, has suf-- fered for years past with severe frontal headache, usually over one eye and in the eyeball. The eyes look red and congested. The headaches came on a few days before menstruation, and hardly ever left her until the second or third day after the discharge had ceased. Menstruation was attended with severe drawing pains all over the body, occasionally darting, quickly changing location, but usually confined to the small of the back. There was considerable tenderness in the spine. She was perfectly regular and the menstrual discharge was usually scanty and dark. The headaches were increasing in intensity and duration. Cured in two months by cimicifuga 3. [Hemple.]

SANGUINARIA CANADENSIS.

Sanguinaria is the remedy, *par excellence*, in sick headache, and particularly in that variety known as "American sick headache." The headache is thus described by Farrington : The patient suffers from rush of blood to the head, and this causes faintness and decided nausea, the nausea even continuing until vomiting sets in. The pains, which are of a violent character, begin in the occipital region, spread thence over the head, and settle over the right eye. They are of a sharp, lancinating character, and at times throbbing. At the height of the disease, the patient can bear neither sounds nor odors. She can not bear any one to walk across the room, for the slightest jar annoys her. As the headache reaches the acme, nausea and vomiting ensue, the vomited matter consisting of food and bile. The patient is forced to remain quiet in a darkened room. The only respite she has is when sleep comes to relieve her. Sometimes the pain is so violent the patient goes out of her mind, or she seeks relief by pressing against her head with her hands or by pressing the head against the pillow.

In migraine or sick headache, the attacks occur paroxysmally ; the pains begin in the morning, increase during the day, and last until evening ; the head seems as if it would burst, or as if the eyes would be pressed out, or the pains are digging, attended with sudden piercing, throbbing lancinations through the brain, involving the forehead and top of the head in particular, and being most severe on the right side, followed by chills, nausea, vomiting of food or bile ; forcing the patient to lie down and preserve the greatest quiet, as every motion aggravates the suffering, which is only relieved by sleep; congestion of the head, with distension of the temporal veins.

Soreness in spots, especially in the temporal region. Headache, as if the forehead would split, with chill and burning in the stomach. Headache begins in the occiput, spreads upward and settles over the right eye ; with nausea and vomiting ; has to be in the dark and lie perfectly still. Pain in back part of head, running in rays from neck upward.

Pain like a flash of lightning on the back of the head. Violent pain over the upper portion of the whole left side of the head, especially in the eye. Attacks occur paroxysmally ; one week or longer. Head is very painful to touch. Headache, with nausea and chilliness, followed by flushes of heat extending from the head to the stomach.

Vertigo, with long continued nausea, debility and headache. Determination of blood to the head, with whizzing in the ears, and transitory feeling of heat ; then a sensation as if vomiting was about to take place. Angry, irritable, morose. Neuralgia in and over right eye. Inflammation of the stomach, with burning, vomiting, and headache.

Case.—Mrs. H., a very fleshy lady of fifty years, nearly passed the climateric, complained of a distressing sick headache hanging about her for years. In some degree the symptoms were almost always present. A typical headache would commence in the forenoon, gathering violence with the hours, until sunset, when it would quietly subside, or else confine her to her bed for a day or two. The pains originated low in the occiput, drawing upward in rays, locating over the right, sometimes the left eye, attended with vomiting, often of bilious matter. She was subject to sudden flushes of heat, burning of the soles of the feet, and that singular symptom, noted in Hale, a quickly diffused transient chill felt at the remotest extremity. At times she had sensible throbbing of every pulse in the body. The urine was generally scanty before and during the severe headache, but quantities of clear urine would pass away when getting better. Prescribed sanguinaria 200, six pellets night and morning for a week. Eight months afterwards, the patient reported relief from the first dose, during the next week complete relief, and from that time until now not a vestige of the old complaint has shown itself, neither the flushes, burning of the soles, electric thrill, or headache. [Dr. J. P. Mills.]

IRIS VERSICOLOR.

Iris is one of the best remedies in sick or bilious headache ; or in headaches beginning with a blur before the eyes. Dull, heavy, or shooting, throbbing pains, mostly in the forehead, accompanied with nausea and vomiting, and great depression of spirits.

Farrington recommends it for sick headaches, particular when they are periodical in their appearance, recurring every Sunday. This is because the strain of the preceding six days has been relieved, and now the patient feels the effect of the strain and has this sick headache. It is especially suited to school teachers, college professors, students, etc. The pain are intense and of a throbbing character, and supra-orbital. They often affect the eyes and cause temporary blindness. At the height of the headache, vomiting often ensues, the vomited matter being bitter, or sour, or both.

Pain dull, heavy, or with throbbing and shooting ; sensation as if there was a band around the head. A tired, aching headache, from mental exhaustion, with violent pains over his eyes, in supra-orbital ridge, occurring on either side, but only on one side at a time. Pain aggravated by rest and on first moving the head, but relieved by continued motion. Headache most severe in afternoon and towards evening.

Constant nausea and vomiting of first watery, sour fluid, then of bile. Sunken eyes with dark, blue rings around them. All paroxysms of pain followed by copious emission of limpid urine and vomiting, with great burning and distress in epigastrium.

Case.—Some time ago, a young lady, aged twenty-seven years, consulted me in regard to a severe headache which had troubled her since she was quite young. The patient, a tall, slender person, of light complexion, flat chest and somewhat stooping, has an attack about once in two weeks, but it is brought on at any time by broken rest, undue excitement, riding in a carriage or boat, and always precedes the menstrual flow. The headache commences with a feeling of prostration and blurred vision and is accompanied by constipation and nausea. The pain is in the left temple and is hard and pressive. It always passes from the left temple to the right between retiring at night and awakening the morning, when it becomes settled in both temples, radiating from both sides to nearly the top of the head. Quiet and sleep relieve ; eating, reading, and excitement aggravate greatly. Iris versicolor, 2d decimal, two drops four times each day, produced a permanent cure, including the constipated habit of the patient. [Dr. I. N. Eldridge.]

GELSEMIUM SEMPERVIRENS.

Passive congestion is the keynote of gelsemium. The cerebral congestion of gelsemium, says Hughes, as also its headache and vertigo, is seen especially in depressed conditions of the system, and causes inability to concentrate the thoughts. Constant, gradually increasing headache, dull, heavy pain, extending to the nape of the neck, frequent throbbing in the temples, and vertigo on rapid movement. Sensation as if a band encircled the temples, going off after copious urination.

Gelsemium is adapted to nervous, excitable, hysterical females ; sensitive people and children ; to male and female onanists. Catarrhal and rheumatic headache when motility is lessened, and the patient is drowsy with neuralgic pains from occiput to forehead and face.

Headache begins in the nape of the neck, passes up over the head, and settles down over the eyes.

It is usually worse in the morning, and is accompanied by stiff neck. The patient cannot think effectively or fix his attention. He becomes listless and stupid ; dizziness with blurred sight and heaviness of the head. These symptoms are relieved by the discharge of pale urine. Headache with soreness of the eyes on moving them. Throbbing in the head which grows worse with the sun.

Headache appears suddenly with dimness of sight or double vision ; with vertigo, great heaviness of head, it feels too big and often too light ; bright red face ; dull, heavy expression of countenance ; full pulse and general malaise.

Heaviness of the head, alleviated on profuse emission of watery urine. Sensation of weight and pressure in the head. Remittent or intermittent headaches. Nervous headache from emotional excitement. Excruciating headache, accompanied by slight nausea ; pain slightly mitigated by shaking the head. Pain most frequently in the forehead and temples. Dull, dragging headache mainly, in the occiput, mastoid or upper cervical region, entending to the shoulders ; relieved when sitting, by reclining the head and shoulders on a high pillow. Nervous headache, the pain commences in the cervical portion of the spinal cord and thence spreads over the whole head. Headache better from shaking the head or bending it forward ; worse about

10 A.M., from anything around the head, when lying down, or from smoking.

Head very light with vertigo. Constant giddiness. Dullness of all mental faculties. Excessive irritability of body and mind. Great heaviness of the eyelids ; great difficulty of opening or keeping them open. Dilatation of the pupil, amaurotic diplopia ; blindness ; dimness of sight. Orbital neuralgia, periodic, every day at the same hour.

Feeling of emptiness and weakness in the stomach and bowels. Diarrhœa from exciting emotions, sometimes involuntary.

Profuse urination with relief of headache. Dysmenorrhœa of a neuralgic or spasmodic character. Feeling as if the heart would stop beating if she did not move about. Pulse full and round, seems to flow under the finger like a current of water, or frequent, soft, weak, almost imperceptible. Disposition to sleep, a sort of stupor. Intense prostration of the whole muscular system.

Case.—A gentleman, forty years old, of nervous temperament, a lawyer by profession and a life-long hard student, had suffered for several years with almost constant headache, characterized by fullness of the head, heaviness in the occipital region, indistinct vision, and occasionally severe vertigo. There exists heaviness in the upper eyelids, great muscular weakness, dullness of mind. During the last six months he has been exceedingly nervous ; can not sleep at night; twitching and jerking. Loss of appetite. Whitish, dirty coating of the tongue. Cured in a short time, by gelsemium 3, three doses each day. [Dr. Hemple.]

IGNATIA AMARA.

Ignatia is indicated in the headache met with in persons of a highly nervous and sensitive temperament, or in those whose nervous system has given way to anxiety, grief, or mental work. The headache is usually situated in one spot in the head, as though a nail were being driven into the spot. Any little mental work or any unusual or severe exertion, any strong odor, pleasant or otherwise, any emotion which would be borne without trouble by one whose nervous system is in a natural state, is sufficient to cause an attack. It often ends in a fit of vomiting. It is often periodical, returning every two days, increases gradually in severity and then suddenly abates. The headaches are characterized by a predominance of pressure ; the pain goes to the eye, which feels as if pressed out ; or to the root of the nose. At the height of the paroxysm, the patient becomes restless and chilly, and often describes a peculiar perversity of vision : she sees fiery zigzags when looking out of the line of vision. Finally, a profuse flow of colorless urine terminates the attack.

The pain is usually semi-lateral ; begins often in

the ear and behind the mastoid process, and runs
up the parietal bone, or back to the occipital protuberance, leaving a stiffness in the nape of the
neck. The pain may also extend from the head
into the eye upon the same side, with burning of
the eye and great sensitiveness to light. Headache
like a pressure from something hard upon the surface of the brain, by paroxysms. Lying down with
the head low first aggravates, then relieves : sitting
up, with the head bent forward, resting on the hand,
is the most comfortable position. When the pain is
at its height the patient tries in vain to shift it by
changing position. Hysteric headache ; headache
from abuse of coffee ; from smoking. At times the
scalp is sensitive.

Mental effort is irksome. Changeable mood, from
sadness to mirthfulness, from crying to laughing.
Great tendency to start. Full of suppressed grief.
Nausea without vomiting ; the nausea when it occurs
follows the pain.

Hering says, Tietzer affirms that ignatia is a
purely spinal remedy. It acts well on sensitive,
hysterical persons, of sanguinous, bilious temperament, delicate, with a tendency to dream day-
dreams, patient cherishing of grief, easily affected
with clonic convulsions from fright or from having
their feelings hurt. The pain is pressing, stitching,
extending from within or outward, especially in the
forehead and in the root of the nose, or as if a nail

were pressed from out of the temples and side of the head. Ignatia helps when there is a pressing pain above the nose, relieved from stooping forward ; pressure from within outward, twitching, beating ; tearing in the forehead, as if a nail were being driven through the head ; stitching, boring, deep in the brain, with nausea ; dimness of vision ; sensitiveness to light ; pallor of countenance ; profuse, watery urine ; the pain often disappears for a time upon a change of position, but returns after eating. Comes on in the evening after going to bed, and in the morning before rising. The patient becomes easily frightened, inconstant or dejected, and disinclined to talk.

PULSATILLA.

Pulsatilla is one of the most valuable remedies in the treatment of various types of headache. These headaches, says Hemple, depend upon a disturbance in some other part of the organism, and may be gastric, bilious, or menstrual headaches. In gastric headaches we will find nausea, vomiting, greasy taste in the mouth, acrid risings from the stomach ; they are usually caused by eating rich, indigestible food, such as pastry, fat pork, etc. In bilious headaches the patient complains of a stupid feeling in the head, and of a sensation in the forehead as if the brain had been bruised. Menstrual headaches will be accompanied by such disturbance of the

sexual system as will direct our attention to pulsatilla.

It is efficacious in headaches from excessive
study, abuse of coffee or spirits, or rich or fat food;
from abuse of quinine, iron, sulphur, mercury,
chamomile tea; from long watching, or cold.
Bland temper, or else cold and phlegmatic. Females and children, with blue eyes, very affection-
ate, easily excited to tears, yielding disposition.
Women who are inclined to be fleshy, with scanty
menstruation.

Acute pains in the temples, with giddiness.
Twitching, tearing pain in the temple lain on ; goes
to the other side when turning on to it ; worse evenings, and on raising the eyes upward. Headache
extending into the eyes, so that they ache in the
evening. Rheumatic headache, worse on one side,
and from 5 to 10 P.M. ; crazing pain in the face and
teeth. Semi-lateral headache, with bad taste in the
mouth in the morning ; without thirst, with nightly
diarrhœa, and scanty urination. One-sided headache, as if the brain would burst and the eyes fall
out of the head. Sick headache, from suppression
of the menses, or from some menstrual or gastric
disorder. Stupefying headache, with running chills,
with humming in the head ; worse lying, or sitting
quietly, or in the cold. Pulsating in the evening,
and from mental exertion ; throbbing with anæmia.

Throbbing, pressing headache, relieved by external pressure.

Giddy as if drunk, with inclination to vomit. Vertigo, especially while sitting, as if caused by intoxication ; in the morning on rising from bed, on account of which it is necessary to lie down again. Can not hold the head upright, can not raise it, it seems too heavy. Disposition to take cold on the head, worse when it gets wet. Head sweaty.

Pulsatilla, according to Hering, cures tearing pains which grow worse toward evening, or beating, lancinating pains early in the morning after rising, or at night after lying down ; it occurs in spells, stitching, tearing in the temple, likely to be one-sided. There is with it vertigo, sickness at the stomach, heaviness in the head, vision becomes somewhat indistinct, and the eyes sensitive to light; roaring, stitching, jerking, tearing in the ears. Pale face, with expression as if the patient were just ready to cry. Loss of thirst and appetite ; shivery and anxious ; occasional bleeding at the nose ; palpitation of the heart. Quiet aggravates all complaints ; the open air relieves them. Headaches better from pressure with hand or bandage.

Pulsatilla is especially indicated when the pain is tearing, beating, lancinating, jerking, chiefly in the temples and occiput, occurring with other characteristics. [*Allg. Hom. Zeitg.*]

Symptoms changeable ; she is very well one hour,

very miserable the next. Timid and fearful; and yet extremely mild, gentle, and yielding. Sometimes silent and melancholy, with thickly coated tongue, bad taste in the mouth in the morning, nothing tastes good.

Case.—A girl, eight years of age, of gentle and sad disposition, has suffered, for six months, with one-sided headache, usually on the left side. The pain is situated in the anterior left temporal region, is beating and stitching, alternately early in the morning after rising and in the evening after retiring, is relieved by external pressure and in the open air; is worse in the room, from lying down or stooping, and from moving the eyes. It continues several hours, increases until it becomes almost unbearable. As the pain decreases, violent gastralgia and sour and bilious vomiting set in. Then pinching, contractive, cramping pains in the bowels. At times these three types of pain appear in alternation; and again every second day. After taking pulsatilla 3 the whole difficulty disappeared, and permanently, by the fifth day. [Pleyel.]

GLONOINE.

The characteristic symptom of glonoine in headache is the throbbing pain, which may be in the forehead, vertex, occiput, in any one part of the head or in the whole head. "This throbbing," as Farrington expresses it, "is not a mere sensation;

it is an actual fact. It really seems that the blood-vessels would burst, so violent is the action of the drug. The throbbing is synchronous with every impulse of the heart. The blood seems to surge in one great current up the spine and into the head. The blood-vessels externally become distended. The external jugulars look like two tortuous cords, the carotids throb violently and are hard, tense, and unyielding to pressure. The face is deep red. This throbbing is either associated with dull, distressing aching or with sharp, violent pains." The headache of glonoine is increased by shaking the head or moving the body, and relieved by external pressure ; but the patient can not bear the head covered.

Congestive, nervous headache with no gastric or bilious symptoms. Violent, throbbing, pulsating headache, with fullness and upward pressure in the head. Congestion to the occiput like a pressure ; such excessive determination to the occiput that it seems as if he would lose his reason. Undulating sensation in the head ; brain feels as if moving in waves ; as if it were expanding itself. Tensive pain over the eyes and nose, also behind the ears, followed by choking sensation about the throat. Headache with red face, accelerated pulse, sweat on the face, unconsciousness; better in the open air, after sleep, after vomiting. Great heat and congestion in the head, especially in the temples and over the eyes, throbbing in the forehead and temples. Hemicrania, sees half light, half dark. Constant inclination to throw the head backward. Severe pain in the occiput extending to the eyes and temples.

Sensation of soreness through the whole head; he is afraid to shake the head, as it seems as if it would drop to pieces. Headache begins with the warm weather, and lasts all summer, increases and decreases every day with the sun; great sensitiveness to the rays of the sun, and to the pressure from covering the head. Sick, faint-like feeling at the stomach, with nausea. Head feels immensely large.

Well-known streets seem strange; way home too long; forgets on which side she lives. Loses senses; sinks down unconscious; congestion alternately to the head and heart. Balancing sensation, constant effort to keep the head erect. Violent action of the heart, distinct pulsation over the whole body.

Cinchona Officinalis.

China, according to Tietzer, is indicated when the pain is a tearing pressure or pressing tearing, especially in the temples, worse at night, disturbing sleep. Aggravation from light touch, walking, draft of air, wind. When the mental powers are stimulated or the imagination roused; in disobedient, quarrelsome persons. It is also indicated in headaches of anæmia, or from loss of vital fluids ; or after intermittent fever.

Hering recommends china in sensitive persons when the pain is pressive and prevents sleep at night, or when there is a pain in the temples as if something were trying to force its way from within outward. Boring in the vertex, the brain feeling bruised and sore. Jerking, tearing, staggering, worse from stepping, walking, every motion, opening the eyes ; better from lying down when everything is quiet. The flesh is sore to touch ; in persons who are dissatisfied ; disobedient and rebellious children, who are fond of sweetmeats, with pale face, only occasionally hot and red, in which case they become very talkative and are very restless at night.

China is indicated, Black says, in drawing, tearing, pressing headaches, which disappear during rest, get worse from slightest motion, and do not remain fixed ; also with passive congestion from loss of blood, active purging, and seminal losses ; with a sensation as from emptiness in the head, ringing in the ears, weakness of vision ; tearing, drawing, crushing pain.

The pain of china is of a tearing, jerking character, usually on one side of the head. Headache, first like a cramp in the vertex, followed by a bruised sensation in side of head, aggravated by the slightest motion. Headache from occiput over

the whole head, from morning until afternoon;
worse lying, must stand or walk, drives to madness.
Headache, head so sensitive it seems as if the skull
would burst ; worse from motion or jar : sensation as if the brain beat in waves against the
skull.

Head much confused in the morning, as after
intoxication, feels weak, can hardly hold it erect.
Whole scalp sensitive to touch, everything hurts
him, roots of hair especially affected. Scalp feels
as if hair were roughly grasped by hand. Sensation
as if the brain were balancing to and fro, and was
striking against the skull, occasioning great pain
and obliging one to move the head. Headache
from suppressed coryza. Headache every other
day. Ringing in the ears and weak fainting
spells.

It is one of our best remedies for intermittent
headaches.

Case.—A girl of fifteen years was attacked with
periodical pains in the head which came on shortly after rising, and continued until the afternoon.
The paroxysms were accompanied by dizziness, and
very violent vomiting. In the evening the patient
was free from pain. The girl became pale, lost her
appetite and strength, and wanted to be lying down
all the time. The urine deposited a brick-
dust sediment. China restored in a very short
period.

SEPIA.

Sepia is in hemicrania one of our principal remedies. It is indicated in pains which localize themselves over one eye, of a throbbing character, deep, stitching pains which seem to be in the membranes of the brain, and which almost always shoot upwards or from within outwards. The patient can bear neither light, noise, nor motion. Associated with this, in women, there is usually soreness of the face and disturbance of the uterine position or of menstruation.

Sepia is useful, Hemple says, in headaches of hysterical persons ; of persons who have been guilty of sexual excesses, whose vitality has become exhausted, and whose countenance looks haggard and worn ; rheumatic and nervous headaches of sensitive, nervous women ; headaches accompanied with chronic indigestion, nausea, constipation, and inability to perform mental labor ; boring, beating headache, usually in the temple or forehead, aggravated by the least touch or motion, and forcing the patient to lie quiet in a dark room ; better from sleeping.

Sepia is adapted, according to Hering, to pressing or boring, also pulsating headaches, chiefly in the temporal region, or under one of the frontal eminences, which often does not bear the least touch. The severity of the pain may give rise to outcries, with nausea and vomiting. Motion is unsupporta-

ble, and usually relief is felt from remaining perfectly quiet and with closed eyes in the dark.
Desire to go to sleep, with ability to do so ; the
sleep affords relief.

The headache occurs in women, and is hysterical ;
in old cases, where the nervous system is out of
sorts. The pain is one-sided, stitching, and leucorrhœa exists between the menstrual periods. If
there is present sudor hystericus, a peculiar sweetish
perspiration in axillæ and soles of feet, sepia is
sure to help. The hemicrania is the outcome of
some affection of the reproductive system ; the
countenance expresses suffering, the features are
distorted. Complexion pale, dirty yellow ; figure
graceful. The pain is mostly stitching, often pressing or boring, tearing, jerking, with passive cerebral
congestion. [Tietzer.]

Stitching, boring, hammering headaches over the
right eye or in one temple, of such severity as to
make her scream, with nausea and vomiting ; better
from sleep and darkness. Hemicrania from affection of the reproductive system ; countenance
pale ; face dirty yellow ; especially in young females in whom the cerebral nerves have excited
the sympathetic, producing a long train of hysterical symptoms. Following the perspiration, headache in the right side of the head and face, not
severe, but with a surging sensation in the forehead,
like waves of pain welling up and beating against

the frontal bone. Pulsating pain in the cerebellum, beginning in the morning and lasting till noon, or sometimes until evening. The pain comes on in terrific shocks, as though there was a powerful jerk in the head. Chronic congestive headaches, with photophobia and impossibility to open the eyes on account of the weight of the upper lid. Headache, with aversion to all food ; better after sufficient sleep. Headache every morning with nausea. Headache all day with great mental depression. Violent headache, as if the head would burst.

Momentary attacks of giddiness when walking in the open air, or while writing. Involuntary jerking of the head backward and forward. Sensitiveness of the roots of the hair, to contact to cold wind ; worse in the evening, when lying on painless side.

Weak memory ; aversion to one's occupation ; great indifference to those one loves best. Easily offended ; inclined to be vehement. Dread of being alone ; great excitability in company. Restless, fidgety, nerves very sensitive to the least noise. Great sadness, and frequent attacks of weeping which she can not suppress. Very sad and fearful about her health.

Affections arise and increase in severity, mostly in the evening and at night ; during and immediately after a meal ; disappear during, or are alleviated by, active exercise, by pressure of the

painful parts, and by application of warmth ; often
accompanied by chilliness.

Case.—A young girl of nineteen years, sanguine-
bilious temperament, somewhat rheumatic, has
suffered for several months with a semi-lateral headache, usually on the right side, which is brought on
by taking even a slight cold, and is tearing, boring
in character, with occasional painful stitches through
the head. If the attacks are unusually severe, the
patient must remain perfectly quiet, close the
eyes, and press the hand firmly upon the painful
spot. Cool air brings on the pains ; it hardly ever
comes on in the night, but may occur at any time of
the day. Can not bear the sight of food ; mushy,
soft evacuations from the bowels. Menstruation
usually a little early, continuing from five to six
days. Sepia 30 cured. [Hirsh.]

ARSENICUM.

The characteristics of the headache of arsenicum
are periodical, burning pains, with restlessness and
anxiety. It is adapted to catarrhal, neuralgic,
and periodic headache, and to migraine in persons
with deep-seated biliary derangements, with vertigo,
nausea, and vomiting of bile. Burning, intermittent
pain having a tendency to periodicity, worse from
cold, with small pulse and cold skin ; the pain is
especially severe on the left side ; can not lean or
rest on that side ; temporarily better from cold

applications. Headaches from the abuse of quinine
and from miasmatic influences. Paroxysms of excessively painful hemicrania, with great weakness
and icy cold feeling in the scalp, followed by itching. Fearfulness; restlessness; dread of death.
Great sensitiveness of the head to open air. Scream-
ing with the pain. Sensation as if warm air were
streaming up the spine into the head. The characteristics of arsenicum are periodicity, adynamia,
great exhaustion after slight exertion ; restlessness
and anxiety ; burning pains, worse at rest and from
cold, and better from warmth and emotion ; thirst,
drinking little but often.

According to Tietzer, it is indicated in hemicrania occurring in persons who give evidence of
plethora abdominalis, where the entire appearance
of patients leads one to suspect liver-trouble, and
bilious colic alternates with migraine.

The pain itself is vague, the patient scarcely ever
being able to describe it with exactness. Usually
the headache is beating, tearing, pressive, stupefying ; is more likely to be confined to the forehead,
with great sensitiveness of the head to the open
air. The patient can find no rest anywhere ; he is
obliged to keep the head and feet constantly " on
the go," fancying this gives him relief. Patient
feels very weak, thinks he is going to die, looks as if
he were bloated, is chilly, seeks the warmth of the

stove. Relief from covering the head with warm clothes and from pressure.

Arsenic, in the treatment of headache, is useful in emaciated persons suffering from affections of the heart or digestive apparatus. Black reports excellent results in a case of a woman who had for twenty years suffered from migraine, attacks of which usually lasted three days, the patient during that time lying unconscious and vomiting violently.

Arsenic causes beating, tearing pain, worse in the warm room, and better in the open air. [Hering.]

ACONITE.

The headache of aconite is characterized by congestion of blood to the head, with fever ; excessive sensibility ; fearfulness ; intolerance of light, noise, or touch; violent, unbearable, stupefying pains, chiefly in the forehead and temples ; nausea and vomiting. The headache follows exposure to cold, dry currents of air ; suppression of perspiration ; anger ; chagrin. The head feels tight and constricted, with fullness and weight in the forehead, sensation as if the brain and eyes would start out, or as if the brain were pushed against the forehead. Crampy sensation behind the orbits, or as if in the bones, or over the root of the nose, with sensation as if one would lose one's reason. Pains are sometimes so violent as to deprive the patient of consciousness,

The extremities feel cold, the pulse is often small and often scarcely perceptible ; the features denote anguish and suffering; the face exhibits a death-like pallor, or else looks bloated, mottled, or dark red. It often relieves the most violent pain when the patient lies as if unconscious, sometimes chokes, moans and complains, thinks he is going to die ; every noise or motion is unbearable, pulse irregular, even intermittent, especially when the headache is of a pulsating, stabbing character, or pinching right above the nose; it is much aggravated from hearing others talk.

In bilious or bilious congestive headaches with many of these symptoms, we may have, in addition, burning distress in the head, as if the brain were moved by boiling water ; or as if the head were en-circled with a red-hot iron. Stupefying pain on the top or on one side of the head, with excessive sen-sitiveness of the scalp, throbbing and stinging pain, as if needles were stuck through the brain ; or the hair feels as if it were standing on end.

Catarrhal headaches with discharge from the nose, roaring in the ears, pains in the bowels ; fever and restlessness; also when there is a sensation as if a bullet were flying around the head, causing a cold draught in it. The headache is better in the open air and worse from talking.

For migraine, with violent stitching, boring pain above the left eye, with nausea, vomiting, worse from a jar, give aconite; in a few hours foll it by sulphur. [Hering.]

According to Black, aconite is adapted to headaches of a congestive or inflammatory nature, when the patient suffers from a sensation of fullness and tension, as if a band were drawn tightly around the head.

Aconite is adapted to sanguine, plethoric persons; and to young and plethoric women in whom the menses are too profuse and protracted.

Hemicrania, with violent pain over left eye, attended by nausea and vomiting. Headache with increased secretion of urine. Fear and anxiety of mind, with great nervous excitability. Gets desperate and declares that he cannot bear the pains. Burning, unquenchable thirst. Bitter, bilious greenish vomiting, with anguish and fear of death. Sensation as if a ball were rising through the brain, spreading a coolness. Headache with arterial tension and excited circulation throughout the body.

CHAMOMILLA.

Chamomilla is indicated in rheumatic hemicrania with tearing, dragging, maddening pains. The tearing, dragging pains which drive the sufferer to despair are characteristic of this remedy. It is also indicated in bilious headaches when the pain is of an oppressive, stupefying, stitching, burning distress, with vomiting of bile, sallow complexion, heavy load and anxiety at the pit of the stomach. Nervous headaches, with violent throbbing in one side of the head, flashes of heat, irritable mood, stinging pains as though the eyes would fall out ; the brain feels sore as if bruised.

Chamomilla seems to act upon a morbidly sensitive nervous system. There is peculiar excitability ; cross and irritable. True of child or adult; can not bear you near them ; answer snappishly ; pain unbearable. Children, light or brown hair; nervous, excitable temperament.

According to Hering, chamomilla often cures if the pain is the result of taking cold or of the excessive use of coffee. Tearing or drawing on one side, extending into the jaw ; heaviness above the nose or very sensitive beating, especially when one cheek is red, the other pale, or the whole face is bloated, with bitter and foul taste in the mouth. Hot sweat about the head, even the hair.

Coma vigil or an inability to open the eyes ; slumbering without sleep ; quick expiration, and tearing headache in the forehead with nausea. Sour smelling sweat during sleep, mostly on the head.

Case.—Mrs. P.,æt. 34 years,suffered from periodically recurring headaches. Attacks often brought on by anger. Violent stitching pains in the left side of the head, extending from the occiput to the left superior maxillary ; sometimes tearing pain in the left ear; left cheek flushed and hot ; frequent shivering, accompanied with fine stitches in the hepatic region. Received at first chamomilla 200, condition unchanged. In the 3d dilution, relief was produced in an hour, and the entire attack soon passed off. [Kallenbach.]

COFFEA.

Hahnemann thus describes the migraine of coffea. It comes on in the morning shortly after waking, and increases little by little. The pain becomes intolerable, and sometimes burning; the integuments of the head are very sensitive, and hurt when touched ever so slightly. Body and mind seem excessively sensitive. The patients look exhausted ; they retire to lonely, dark places, and close their eyes in order to avoid the light of day ; they remain seated in an arm-chair, or stretched upon a bed. The least noise or motion excites the pain. They avoid talking or being talked to, or hearing others talk. The body is colder than usual, although no chills are experienced; the hands and feet especially are very cold. They loathe all food and drink, on account of continual sickness at the stomach. If the attack is very violent, a vomiting of mucus takes place, which, however, does not diminish the headache. There are no alvine discharges. The attacks come on irregularly and without premonitory symptoms.

It is adapted to nervous hysterical headaches, in

people with nervous or sanguine temperaments. Headache from excessive joy ; from meditation, contradiction, chill, etc., especially in patients who do not habitually use coffee, or to whom it is repugnant. According to Tietzer, it is adapted to migraine when the pain is drawing, pressing, as if a nail were pressed into the side of the head, or as if the entire brain was crushed. The pain drives the patient to despair ; he runs about in the room, dreads noise, shivers, is afraid of the open air ; it is worse after eating. If a special mental effort causes migraine in a person not accustomed to coffee, the remedy is quite sure to cure.

One-sided headache, as if a nail were driven into parietal bones, or as if the whole brain were torn and bruised; worse in open air. Crackling in the vertex when sitting quietly. Head feels too small.

Hering recommends coffee in violent, one-sided, drawing, pressing pains, as if an nail were driven into the side of the head, or as if the brain were crushed or torn; it often appears from trifling causes, close thinking, anger, taking cold, over-eating, etc., with aversion to coffee, which usually tastes good, sensitiveness to noise, music ; the pains seem perfectly unbearable, giving rise to whining; the patient is actually beside himself, screams, cries, throws himself, has great distress, dreads the open air, is chilly.

NATRIUM MURIATICUM.

This remedy, although having quite an extensive range, is not as frequently used in the treatment of headache as it should be.   It is applicable to cachetic persons, and to those who have lost animal fluids. It is adapted to chronic and to sick headaches ; to headaches before, during, and after the menses ; to cephalalgia of school-girls who apply themselves too closely to their lessons ; to headaches commencing in the morning after waking, lasting until noon, or going off with the sun ; to catarrhal headache, and to migraine.

The headache of natrium muriaticum is worse from any mental exertion.   In the morning on awakening, there is throbbing, mostly in the forehead, as if from many little hammers beating in the head. The pain is so severe at times as to make the patient almost maniacal. With this kind of headache, the tongue is dry and almost clings to the roof of the mouth, although it may look moist when it is put out. There is great thirst. The pulse is almost always intermittent.

Hahnemann recommends natrium for daily feeling of heaviness in the head, especially in the occiput, drawing the eyelids together. Pressure throughout the head and in the temples ; headache early in the morning upon awakening, as if the head would burst ; stitches and pressure above the eyes. Beating, hammering, and pounding in the head.

Shooting from the forehead to occiput, then in a minute the reverse ; feels giddy as if she would fall to the left ; room seems to revolve ; eyes close with pain ; better in open air.

Headache in the vertex and occiput like a weight, worse in the evening, relieved by pressure, with heat of the face. Right-sided headache, comes on at 10 A. M., with dizziness and dull, heavy pains, glimmering before the eyes, fainting and sinking at the epigastrium, slight fever and thirst ; better in the open air, and while perspiring ; sore nostrils. Pains in and over right eye, going off with the sun; can not bear light of any kind. Dull, stupefying headache after waking in the morning, lasting till toward noon. Pain like a nail driven into the left side of the head. Pressure pain in the forehead and eyeballs so violent that the lids can only be raised with exertion and pain.

Cold sensation on the vertex and painful sensitiveness of the scalp, with spasms of the eyelids. Feeling of heaviness in the head as if it would fall forward, increasing with headache across the forehead (as from a blow). Vertigo and great dullness of the head, with flickering before the eyes. Almost constant dull headache. Headache from sunrise to sunset ; worse at midday ; right eye congested ; worse from light. Headache wakens at 12 P.M., lasting until 10 A.M. Periodic vertigo with nausea, eructations, headache sometimes as if a cold wind were blowing through the head. Constant heat in the afternoon, with violent headache and unconsciousness gradually relieved by perspiration. Nausea immediately after eating, with heaviness of the head and bitter eructations. Aversion to bread, craves salt. Constant chilliness and want of animal heat. Sad and weeping; consolation aggravates, with palpitation of the heart. Gets angry at trifles; hateful, vindictive, joyless, indifferent, taciturn; likes to dwell on past occurrences.

Case.—Mrs. P., seamstress, has a sick headache whenever she eats rich food. The attack commenced with dazzling in the eyes, like lightning, which lasted half an hour and ushered in a throbbing headache in the forehead and vertex, with nausea; she can hardly hold her eyes open; the feet are cold, and there is great chilliness all over; occasional sour or bilious vomiting. Prescribed natrium muriaticum 200, every three hours. The first dose cured the attack. [Dr. J. C. Morgan.]

CALCAREA CARBONICA.

According to Tietzer, calcarea carb. is next to sepia in the treatment of migraine, especially in persons who are pre-disposed to affections of the reproductive system and are of scrofulous habit. The pain is of a dull pressure in the vertex, running into

CALCAREA CARBONICA. 67

the forehead, or a drawing pain in the forehead, with coldness in it, and nausea, brought on or aggravated in the open air. It is especially adapted to leuco-phlegmatic temperaments, and is characterized by anxiety, fearfulness, melancholy, and difficulty in thinking, dilated pupils and dimness of sight. The pains are felt chiefly in the forehead, on the vertex, or on either side of the head. It is useful in menstrual headaches, with pallor of the face and coldness in and on the head ; empty eructations, nausea, profuse menstruation. Feet cold as if they had on damp stockings.

Chronic cases after suppressed eruptions ; strange feeling of coldness in some parts of the head, or in the whole head ; pain worse from early in the morning after getting awake until afternoon ; sweaty hands and feet. Stupefying, aching, beating or hammering pains, or hemicrania, with nausea, eructations, and desire to lie down ; or boring in the forehead as if the head would split ; falling off of the hair.

Frequent one-sided headaches, always with empty eructations. Headache beginning in the occiput and spreading to the sinciput, so severe she thinks the head will burst. Chronic headache depending on brain-fag.

Relief from tight bandaging, or from cold applications.

Black says in its effects calcarea resembles belladonna, especially where complaint is made of heaviness, fullness, pressure, and heat in the face. Violent pressing pains above the eyes, with trembling of the lower eyelids, strongly call for its use, especially so when the patient awakens in the morning stupid and unrefreshed, and there is a tendency to nervousness.

Weight and pressure in the forehead, obliging him to keep his eyes closed ; headache from reading and writing ; boring in the forehead as if the head would burst ; pulsating pain in the occiput ; throbbing in the centre of the brain ; hammering headache after walking in the open air, forcing him to lie down ; headache and buzzing in the head, with heat in the cheeks ; icy coldness in the right side of the head ; nightly sweating of the head [Hahnemann.]

Tearing pain in the right frontal region with great sensitiveness to touch. The headache usually involves the right side and seems to spread in streaks.

Case.—A young girl, age 21 years, of graceful figure, quiet, sensitive, who menstruates every three weeks, with preceding swelling and tenderness of the breasts and leucorrhœa, suffered for several weeks from beating, pulsating frontal headache, with internal heat of the head and feeling as if it would burst ; this begins every afternoon, growing worse in the evening. The blood rushes to the face, the cheeks are scarlet, spotted ; when sleeping, noisy almost snoring breathing. Permanently cured by calcarea carb. 30. [Hempel.]

SULPHUR.

Sulphur will prove useful in congestive, gastric, nervous, or catarrhal headaches, or headaches from suppressed eruptions, from abuse of spirits or metallic substances ; when characterized by pressing, throbbing, or tearing pains mostly in the forehead and temples. Constant heat on top of the head and coldness of the feet.

In beating, tearing headaches, with heat in the morning or at night, with nausea, worse in the open air, better in the house ; tearing with stupefaction, pressure ; occur weekly with falling out of the hair, suppressed eruptions and boils.

Sulphur is indicated, Black says, in headache which is located in the forehead and vertex, when the patient complains of hot head and cold feet, the head is pressed together, and there is hammering and ringing in the ears ; further, in congestive states, affecting the head, in individuals of sedentary habits, suffering from constipation and hemorrhoidal, who have had old eruptions, etc., suppressed, or in whom the hemorrhoidal discharge has been stopped.

Hahnemann recommends sulphur for nightly headaches upon the slightest motion in bed.

Heaviness in the head, especially in the occiput. Each day drawing headache, as if the head would burst. Stitching, buzzing headache. Beating headache in the vertex, coldness ; a cold spot in the head.

Sulphur will also prove a useful remedy in chronic headache occurring in paroxysms, with the characteristic symptoms. Pressure and tight feeling in the brain ; tearing or boring pains in the forehead and temples.

Painful tingling in the temples and in the vertex. Sensitiveness of the vertex ; pressing pain when touched ; worse in the evening ; from heat of bed ; in the morning when awaking ; smarting and burning after scratching. Throbbing headache at night. Periodical headaches every seventh day. Sick headache, very weakening, once a week or every two weeks ; pains generally lacerating, stupefying, benumbing.

Headache from abdominal plethora, from suppressed skin diseases ; chronic gouty and rheumatic headaches ; increased by motion, mental exertion, etc. Dull headache, commencing in the morning, increasing till noon or a little later, and then gradually decreasing.

Contractive pain, as from a band around the cranium, with sensation as if the flesh were loose ; followed by inflammation, swelling and caries of the bones ; worse in wet, cold weather, and when

at rest ; better from motion. Aching pain over
the eyes, obliging one to knit the brows or close
the eyes ; or headache with dim sight, inability to
think ; nausea and desire to vomit. Vertigo,—while
sitting or standing, with nose-bleed, in the morning;
when stooping ; when rising from bed ; when walking in the open air ; with nausea ; vanishing of
sight ; with inclination to fall to the left side ;
worse after meals, particularly dinner. Roots of
hair painful, especially to touch. Hair dry, falling
off, scalp sore to touch, itching violently in the
evening, when getting warm in bed.

Weak memory, especially for names. Hypochon-
driac mood during the day, merry in the evening.
Excitable mood ; easily irritated, but quickly peni-
tent. Melancholy mood ; dwelling on religious
and philosophical speculations. Frequent weak,
faint spells, through the day.

Case.—A young lady had inherited from her
father a cephalalgia which appeared two or three
times a week. During the remainder of the time
she was in good health ; no digestive disturbance.
Her father had been vaccinated with vaccine taken
from a scrofulous infant affected with a cutaneous
eruption. Some time afterward he was taken with
a cephalalgia which he retained to his death. The
young lady received sulphur three times a day.
The cure was complete in about two months. [Dr.
Gaudy.]

SPIGELIA.

Spigelia is indicated in congestive, nervous, or rheumatic headaches; particularly those resulting from, or coincident with, affections of the eyes or heart.

The type is characteristic. The pain occurs in the head, eyes, face and teeth, more often in the morning until noon; motion, and especially stooping forward, aggravates the pain. One of the keynotes is "pressure in the head from within outward" and "sensation as if the contents of the cranium were trying to crowd out through the forehead," both of which are worse from stooping. [Hering.]

Sharp neuralgic pains over the left eye; the pain comes up from the nape of the neck, and over the head, then settling over the left eye. The pain begins in the morning with the sun, and increases until noon, when it reaches its height, and then gradually decreases until sunset, when it ceases. The suffering is increased by any noise or jarring of the body. The pains are usually worse in changing weather, particularly stormy weather. At the height of the pain there is often bilious vomiting.

Pains darting from behind forward through the eyeball, causing violent pulsating pains in left temple and over left eye. Painfulness of the cerebellum with stiff neck. Nervous headaches, especially when the eyes are involved ; worse from thinking, noise, or any jarring ; pale face, anxious respiration, palpitation, nausea and vomiting. Any quick movement converts the dull, aching pains into acute stabbing. When moving facial muscles, sensation as if the skull would split.

Headache from stooping, as from a band around the head. Neuralgia, where the pain centers on the eye, or above or below, from cold, in damp, rainy weather.

Tietzer commends it for left-sided hemicrania, involving the entire side, both head and teeth. The pain is digging, tearing and heavily pressing, worse from stopping, from the slightest motion in the open air, from each loud noise. Tendency to gouty affections in other parts of the body.

It is also useful, according to Black, in nervous rheumatic headaches, with heavy pressing pain extending into the face and neck, worse from touch, motion, and noise.

Farrington gives as a characteristic symptom, sensation as if the head were open along the vertex.

Case.—A man, twenty-five years old, of sanguine-bilious temperament, of strong constitution, had an attack of headache three years ago, which was repeated two years ago, and again this year. Symptoms, continuous violent pain, at times jerking, and lacerating as if fine instruments were lacerating the nerves ; this attack is located in the right forehead and temple, but affects also the right eye and the superior maxillary. The right eye is quite inflamed ; there is considerable lachrymation, and a feeling as if the eye was being forced out of the socket, with sensitiveness to light. Burning heat in the affected parts and in the face. The temporal arteries pulsate strongly. There is coryza, restless sleep. Belladonna and others did no good. The pain became perfectly excruciating and communicated itself to the teeth. Prescribed spigelia 30. After half an hour, violent perspiration, which continued through the night ; he found himself well in the morning. [Dr. Romig.]

SILICIA.

Silicia is adapted to congestive, gastric, nervous and rheumatic headaches. There are beating, pulsative pains, with heat and rush of blood to the head, excited by mental effort, talking ; nightly pains extending from the nape of the neck to the vertex ; neuralgic tearing pains every afternoon ; sensation as if the head would split and as if the brain would issue through the forehead and eyes ; semi-lateral stitching or tearing pains, extending to the nose and face ; tumors on the head, the hair falls out, and the skin becomes very sensitive.

According to Farrington the headache is of a

nervous character. It is provoked by any excessive mental exertion. Then it is usually supra-orbital, and is generally worse over the right eye. It is worse from any noise, motion or concussion, and better from wrapping the head up warmly. It is not the pressure, but the warmth, that relieves. Sharp, tearing pains rise from the spine into the head. At the height of the paroxysm there is apt to be nausea from involvement of the stomach.

Headaches from excessive mental exertion, from over-heating, from nervous exhaustion. Scrofulous diathesis. Rachitic, anæmic conditions ; caries. Nervous, irritable persons, with dry skin, profuse saliva, diarrhœa, night sweats. Weakly persons, fine skin, pale face, light complexions, lax muscles. Persons who are over-sensitive, imperfectly nourished, not from want of food but from imperfect assimilation.

It is recommended by Hahnemann, in headache from the neck to the vertex, preventing sleep at night ; daily headaches, tearing in the forenoon, with heat in the forehead ; from noon until evening, a heaviness which becomes a crowding through the forehead ; bursting pain ; sweating of the head in the evening.

Black says, headache due to organic causes or to excessive study, or to nervous exhaustion depending upon any other cause, with vertigo and weakness of memory,

Vibratory sensation in the head when stepping hard, with tension in the forehead and eyes. Hemicrania, with loud cries, nausea to fainting, subsequent obscuration of sight.

Severe pressing or shattering headache, the pain is felt in the nape of the neck, ascends to vertex, and then to supra-orbital region ; also from the occiput to the eyeball, especially the right one, sharp, darting pains, and a steady ache, the eyeball being sore and painful when revolving ; worse from noise, motion, even the jarring of the room by a footstep, and from light ; relief from heat but not from pressure. Violent periodic headache, vertex occiput or forehead ; one-sided, as if beaten ; throbbing in forehead ; coming on in night, with nausea and vomiting. Periodical headache every seventh day. Obstinate morning headache, with chilliness and nausea. Violent headache with loss of reason. Headache, with, or followed by, severe pain in the small of the back ; heaviness and uncomfortable feeling in all the limbs.

Vertigo ascending from the dorsal region through nape of the neck into the head ; worse from motion, or looking upward ; accompanied by nausea. Congestion to the head ; cheeks hot ; slight burning in the soles of the feet. Cold feeling from nape of neck to vertex, extreme heaviness of head.

When crossed, has to restrain himself to keep from doing violence. Confusion of mind; difficulty of fixing attention. Over-anxious about himself; low spirited ; desponding ; tired of life ; indifferent ; apathetic. Reading and writing fatigue, can not bear the least thought. Hungry but can not get food down. Water tastes badly, vomits after drinking. Susceptible to slight draughts of air. Want of vital warmth, even when taking exercise.

ARNICA MONTANA.

Arnica is not a prominent remedy in headache, although in the rare cases in which it is indicated it will do good service. Its sphere of action is more in headaches of arthritic or rheumatic origin, although King recommends it in congestive and gastric headaches. In headaches from mechanical injuries of all kinds, from a fall, a blow, or any similar cause, arnica is usually efficacious. The pains are aching, darting and pressive, mostly in the forehead ; or over the eyes, worse from motion. Head and face hot, while the body is cool. Pain as if a knife were drawn through the head, transversely from the left side, followed immediately by internal coldness of the head. Headache as if a nail had been thrust into the temple, with general sweat about midnight, followed by faintness. Burning in the brain, with natural heat of the body, night and morning ; worse from motion, better from rest. Violent, stupefying headache on waking in the morning. Feeling of cold at small places on the forehead, as if touched with something cold. Soreness in the stomach, and eructations tasting like putrid eggs. Nausea and vomiting worse after eating and drinking. A warm room is unbearable, but the open air does not relieve.

Case.—A man, aged 35 years, received a knock upon his head, and suffered from the following symptoms, which grew worse steadily : Pressing pain in the forehead ; painful, dull pressure upon the margin of the right orbit; heat in the head while the rest of the body was cool ; aggravation of the pain while eating ; heat in the face ; roaring in the ears ; contracted pupils ; nausea in the morning ; dislike to tobacco-smoke ; desire to go to stool frequent, but the effort is ineffectual ; anxious dreams ; considerable fever in the late in the evening ; complaining, disagreeable mood. After arnica, 6th dil., all the symptoms disappeared within two days. [Baudis.]

FERRUM.

Ferrum is recommended by Hughes in the headaches of passive congestion, and in the pseudo-hyperæmic headaches following large losses of blood. Pulsating pain in the head, worse usually after midnight. The face is fiery red during the attack, and the feet are cold. Vertigo, which is worse from rising suddenly from a lying to a sitting position, or it is brought on by crossing a bridge, or by running water, or riding in a car or carriage.

Headaches depending upon gastric difficulties, with nausea, pressure in the stomach after eating. Headaches of chlorotic patients. The pain drives the patient out of bed, must walk about. Pressive headache in the forehead as if it would burst. Hammering and throbbing headache. Violent shooting headache, in the left side, in the afternoon. Drawing from the nape of the neck into the head, with stinging, hammering, and roaring in the head. Headache as if the brain were rent asunder. Pain in the head, as from subcutaneous ulceration, and painfulness of the hair when touching it. Headache every evening above the root of the nose. Every two or three weeks, headache for three or four days, hammering and beating so that she must sometimes lie down in bed.

Case.—Mrs. E. F. Severe headache, pain (following a uterine hemorrhage) so great that she can not lie still, yet is too weak to move about. Painful beating as from many little hammers. Face, hot, red. Ferr. 30 cured at once.

CAMPHORA.

The range of camphor in headache is quite limited, but it is the only remedy offering the symptom of beating pain in the cerebellum, synchronous with the heart-beat. The head is hot, face red, limbs cool, better on standing ; mostly with such as were deprived of their usual sexual intercourse.

Frontal headache; pressing outward; also left-sided. Contraction, as if laced together in the cerebellum and globella, with coldness all over. Pains run from head to tips of fingers, with trembling and nervousness.

Great anxiety and extreme restlessness. Sudden and complete prostration of the vital forces, with great coldness of the external surface. Extremities cold and blue with cramps. Impotence, with coldness, weakness, and atrophied condition of sexual organs.

Case.—A woman, æt. 19 years, had pulsating, beating in cerebellum, synchronous with the pulse beat. Camphor, one drop every ten minutes, cured her in a few hours. (*Med. Counsellor.*)

Cocculus.

Cocculus indicus is homœopathic to congestive, gastric, and reflex uterine headache, and is of great value in many sick headaches. It is especially suited to women and nervous children, of a lively turn of mind, troubled with imaginary fears. A characteristic symptom of cocculus in headache is intense pain in the occiput, as if the parts were alternately opening and closing. Violent headache; unable to lie on the back of the head; is forced to lie on the side; unable to bear the least noise; noise excites vomiting. Rueckert recommend cocculus for headache worse from eating and drinking and accompanied with a feeling of emptiness and hollowness in the head, for which the patient could not find expression.

Headache with nausea and inclination to vomit. Hughes recommends it for headache with nausea resembling sea-sickness, as if the stomach heaved up and down. It is therefore useful in headache from riding in a carriage or a boat. Tearing, throbbing headache especially in the evening. Pressing headache from within outward; or as if compressed by a bandage; or as if screwed together, Headache, as if something forcibly closed the eyes. Violent headache which compels the patient to sit up, aggravated by talking, laughing, noise or a bright light. Headache as if the eyes would be torn out, particularly during every motion, with vertigo.

Cocculus, according to Black, agrees with hypochondriacal, melancholic persons, inclined to be worried and frightened. In nervo-gastric headaches with sickness at the stomach or actual nausea, or in headaches during the menstrual flow, especially when it appears too soon.

There may be also confusion of the head, most increased by eating and drinking. Cramp in the stomach after eating and drinking, with oppressed breathing. Excessive hunger. Repugnance to food, accompanied by hunger. Dysmenorrhœa, always followed by hemorrhoids. Intolerance to both cold and warm air, although the headache is ameliorated by quiet and warmth, and extremely aggravated by cold air.

The condition of the nervous system set up by menstruation and pregnancy is extremely favorable to the action of cocculus.

Case.—Miss H., æt. 35, of a full plethoric habit, has suffered from her present headaches for now fifteen years ; they came on shortly after the catamenia appeared, and have ever since regularly occurred at that period. Violent headache—described as a dull pain affecting the whole head ; the patient has a difficulty in describing it minutely ; is unable to lie for a moment on the back of the head ; is forced to lie on the side ; unable to bear the least light ; any noise excites nausea and vomiting. During the headache she feels as if suffering from sea-sickness, and, on sitting up, the objects around seem to move up and down. The headache lasts from thirty-six to forty-eight hours, and comes on the third or fourth day of the catamenial period. The catamenia are abundant, but unattended by local pain. General health good. She was treated from March 16 to May 1, with cocculus 18, night and morning. Had no return of the headache after October.

AGARICUS MUSCARIUS.

This remedy is indicated in headaches occurring in those who are subject to chorea, or who readily become delirious in fever or with pain. It is also adapted to nervous and congestive headaches, where there is pressure, sleepiness, frequent and violent yawning, muscular relaxation and weariness, soreness over the entire body, backache, and a sensation as if the joints were out of place. Violent oppressive pains. Pains as though sharp ice touched the head or cold needles pierced it ; icy coldness in region of the coronal suture, or sensation of coldness on right side of frontal bone, though warm to touch. Drawing from both sides of frontal bone as far as the root of the nose. Pain as from a nail in the right side of the forehead. Confusion of the head, heaviness, as after intoxication. Vertigo, brought on by protracted mental application, or exciting debates.

Frequent slight twitching of the eyelids, or of the facial muscles. Violent, shooting, burning pains deep in the spine. Painfulness along spinal column when stooping. Spinal column sensitive to touch, worse in the morning. Sensation as if ants were creeping along the spine. Frequent jumping of muscles in different parts of the body ; general trembling.

Asarum Europæum.

Depression of the cerebral function with heavy headache is, according to Hughes, the characteristic of this remedy. There is an inability to think, a mental dullness, with giddiness and a feeling of intoxication. Stupid feeling in the head ; has no

desire to do anything. The headache is of a rheumatic nature, drawing and tearing; throbbing pain
in the forehead, sides of the head; the drawing
and tearing pain, in some cases, reaches downward
to the root of the nose, or to the nape of the neck.
The headache is unaccompanied by nausea, and
abates after vomiting. Intense, compressive headache, more violent when walking or shaking the
head. Tension of the scalp; the hair feels painful.
Cold feeling at a small spot on the left side of the
head, a few inches above the ears. It is suitable
for nervous temperaments, excitable or melancholic
mood.

Case.—A young lady of good constitution has
suffered for several years from headaches. Symptoms: Keen pain over the left eye, with lachrymation from the affected eye, can not use the eye to
read; symptoms aggravated by a bright light; the
temples are painful externally, can not dress the
hair on account of it; if the attack is violent, both
eyes become painful and she has nausea; constipation; menstruation normal. The attacks frequently precede or follow menstruation. Asarum 1-200,
given during the intervals, cured. [Dr. Gross.]

AURUM METALLICUM.

The key-note of gold is mental weariness and depression, with thoughts of suicide: there is anguish
of the mind; grief, hopelessness, and despondency.

Disposition to grumble and quarrel and to fly into a passion. The pain is stitching, or a hard, aching, bruising pain. Bruised pains, especially early in the morning, or during mental labor, so that the ideas frequently become confused ; roaring in the head, in hysteric females; the latter symptom is un-doubtedly due to disturbances in the uterine sphere. [Hartmann.]

The headache to which gold is more frequently adapted is the congestive variety ; congestion of blood to the head, the brain feels sore and as if it had been bruised; semi-lateral, acutely throbbing headaches ; pains in the bones of the head on lying down, as if they were broken. Fine tearing from the side of the head, through the brain, violent dur-ing motion. Tearing headache in front, in the fore-head, in the vertex deep in the brain, abating in the open air. Megrim returning every three or four days, with stitching, burning and beating in one side of the forehead, with nausea, and even bilious vomiting. Headache as from incipient cold. Severe and constant heat on top of the head. Prick-ings as from pins in the forehead externally. Sen-sation as if a current of air were rushing through the head, if it is not kept warm.

Headaches of persons with black hair, dark olive-brown complexions, disposed to constipation, sad, and gloomy, taciturn ; or to sanguine individuals, with black hair and black eyes, a lively, restless, anxious disposition, always disposed to feel anxious for the future.

Case.—M., twenty-two years old, has suffered for two months from severe left-sided hemicrania; violent cutting, stitching, and tearing pains which change from the left forehead to the left teeth, are quiet for a short time only, and return without provocation. They disturb his sleep. Decayed teeth. Mental depression and weariness of life, with thoughts of suicide. Weakness of the legs, especially of the knee-joints, oppression of the chest, stitches at times during inspiration, dart into the back. Pollutions, masturbation; weakness of memory. Prescribed aurum 2, twice each day. Cure of the pains on the second day, and early, general improvement. [Schleicher.]

COLOCYNTH.

The headaches to which colocynth is particularly adapted are those of a neuralgic or gouty diathesis, often of an intermittent character, or confined to one side of the head. The pains are screwing, as if the head were in a vice; or the pains are stitching, tearing, and digging, and the eye is sympathetically affected. One-sided headaches with nausea and vomiting. Severe headache with sweat on the head and body, smelling like urine, with copious watery urine during the pains, or scanty, fetid urine between the paroxysms.

Colocynthis helps, Hering says, when the pain is
of the greatest intensity, savagely tearing, or one-
sided, drawing, pressive. Pressure in the forehead,
made worse from stooping forward or lying on the
back. Attacks appear every afternoon, or toward
evening, in the left side, with great restlessness and
anxiety ; especially when the sweat has a urinous
odor, when the urinary secretion is scanty and very
offensive, but is copious.and quite clear during the
paroxysms. .

According to Tietzer, colocynth is the remedy
when the pain is one-sided, pressing, squeezing, or
drawing, with nausea and vomiting. It has a tendency to extend into the forehead and left side,
and usually appears toward evening. It is worse
from stooping forward, motion, shaking of the head,
and moving of the eyelids. Emotions bring on the
pain, especially indignation, anger or mortification
on account of ill-treatment. There is often found
connected with it a great deal of restlessness and
anxiety.

The hemicrania which comes within the reach of
colocynth is based upon an exaltation of sensibility, caused by rheumatic or gastric disorders.
Hemicrania which finds its cause in the thickening
of the arachnoid, in growths in the cerebral covering, or other organic changes, lies beyond its reach.
Usually the hemicrania of colocynth lies in the
course of the frontalis, is accompanied by severe

pains in the eyes, and alternates with neuralgias of
the plexus cœlicacus. [*Oestr. Zeitschr.*]

Case.—A boy of thirteen years had been complaining for four days past of violent stitches in the
forehead and eyes, darting from without inward.
The pain continued day and night, abating only momentarily, and returning all the more violently after
an abatement. The boy had fever, a bitter taste in
the mouth, complete loss of appetite, and constipation. Six hours after one dose of colocynth 30, the
pain disappeared, and on the day following the patient left his bed. [Dr. Attomeyer.]

CROCUS SATIVA.

Crocus seems more particularly adapted to headache at the change of life, most severe at the time
when the menses used to occur, persisting interrupt-
edly for two or three days, having but few inter-
missions, even during the night, sufficiently long to
allow of sleep.

Transient, broad thrust above the left frontal
eminence, extending deep into the brain, causing
him to start ; a painful confusion remains for a
moment, relieved by external pressure. Pulsating
pains in one side of the head extending into the eye.
Rhythmical pulsations of the whole half of the head
and face. Constant, sensitive pain as if a dull point
were pressing inward ; succeeded by intermittent
paroxysms of the same pain. Pain in the forehead

in the evening, by candle-light, with burning and pressure in the eyes. Sensitive tearing in the head and right eye, with dimness before the left eye and sensation as though a draft of air blew into it. Headache above the eyes with pressure and burning in them, which compels rubbing, and is much increased toward evening, especially in the right.

Sensation, when moving the head, as though the brain were loose and went tottering to and fro.

Changeable disposition ; depression and hilarity, or ill-humor and lively mood, alternating. Sings involuntarily on hearing even a single note sung ; laughs at herself ; but soon sings again in spite of her determination to stop. Gayety, uncommon mirth and cheerfulness ; witty, jocose, loquacious. Great forgetfulness, loses his ideas when he undertakes to write them down. Takes everything in anger, and suddenly repents of having injured others. Disagreeable mood ; vehement, peevish, quarrelsome ; an hour later talkative, laughing, singing.

Must wink and wipe the eyes frequently, as though a film of mucus were over them. The light seems dimmer than usual, as though obscured by a veil. Feeling as though water was constantly coming into the eyes, only in the room, not in open air. Epistaxis of very tenacious, thick, black blood, with cold sweat on the forehead in large drops. Excessive prostration and weariness in the evening, as from severe physical exertion, accompanied by great sleepiness, with feeling as if the eyelids were swelling ; literary occupation relieves.

Case.—Crocus 3, two or three doses daily, and repeated whenever a recurrence of the attack took place, cured those headaches which frequently occur during the climateric period, of a jerking, throbbing character, now here, now there, with congested blood-vessels, not only in the head but in other parts of the body, and pressure upon the eyes. It was worse at the time when formerly menstruation would have taken place, then continuing from two to three days, stopping for a short rest. [*Allg. Hom. Zeitg.*]

DULCAMARA.

Dulcamara is especially adapted to catarrhal and rheumatic headaches, in damp, cold weather. The symptoms are all aggravated when the weather suddenly becomes colder, especially if it be damp. Headache of a stupefying character, with coldness of the whole body, and disposition to vomit ; boring, or digging as if the brain would expand ; or as if a board were pressing against the forehead. Continuous dull, heavy headache. Stupefying pain in the occiput, ascending from the nape of the neck. Oppressive, stupefying pain in forehead, with obstruction of the nose, aggravation during motion, with heaviness of the head.

Chillness in cerebellum and over the back, returning every evening ; cerebellum and whole head feels as if enlarged ; worse in cold, damp weather ; worse until 2 P.M. Mental confusion, can not concentrate his thoughts. Phlegmatic, torpid, scrofulous patients, who are restless and irritable ; who take cold in changes. Increased secretion in mucous membranes and glands, those of the skin being suppressed.

Case.—F. P., aged twenty-seven years, has complained for two years of a continous, dull pain in his head, chest, and stomach, attended with great uneasiness, depression of spirits, labored respiration with mental confusion and inability to collect his thoughts. For a year he had been treated for torpid liver and impaired digestion. Phosphoric acid, third dilution in water, seemed to relieve the case at first, but soon ceased to give the patient any relief. At last he received dulcamara, the third centesimal, in water, twice a day for a week, the administration of the remedy being followed by a complete cure. [Dr. Small.]

LACHESIS.

The sphere of lachesis is quite extensive, as it will be found useful in almost all varieties of headaches ; catarrhal, congestive, hysterical, rheumatic and menstrual. It is especially suited to headaches accompanying menstrual irregularities, and to women

at the climateric period. It is characterized by
nausea, and drowsiness, throbbing or beating pains
in the temples. Pressing headache early in the
morning, worse from stooping. Can not bear anything tight around the waist. Vertigo with paleness
of the face. Pain in the left ovarian region. Lar-
ynx and throat very sensitive to touch. Despon-
dent mood. Aggravation after sleeping.

Headache over the eyes and in the occiput every
morning on rising. Aching under the skull all over.
One-sided headache ; pain intense, extends to the
neck and shoulders ; neck stiff and tongue para-
lyzed. Headache after a cold ; first a pressure in the
forehead, which increases to a violent beating in the
evening, with nausea and inclination to vomit. Painful sensitiveness of the whole left side of the head.
Giddiness, with headache, particularly before the
menses.

Farrington says of lachesis, headache worse in
or over the left eye, of a throbbing character, with
sharp pains, very severe, which may come at the
climaxis, or as the accompaniment of an ordinary
cold ; relieved as soon as the coryza appears.
There is a universal characteristic : so soon as a
discharge is established the patient feels better.

According to Black, lachesis is indicated in one-
sided, tensive headaches, extending from the occiput to the eyes, with stiffness of the neck and sensitiveness of the scalp.

MERCURIUS SOLUBILIS.

The headache to which mercurius is adapted is principally of a catarrhal, rheumatic, or bilious nature. Feeling of fullness if the skull would split, or as if the head were tied up with a bandage ; tearing, burning, or stitching and boring pains, or semi-lateral tearing down to the teeth and neck, with stitches in the ears ; violent aggravation at night, by warmth of the bed, also by contact, hot and cold things ; constant night sweat, but without relief.

Bilious congestive headache. Violent aching pain in the whole head, with feeling as if the brain were sore ; accompanied by a copious flow of water from the mouth, vomiting of green and yellow bile. Head feels as in a vise, with nausea ; worse in the open air, from sleeping, eating and drinking ; better in the room. Head feels as if it was getting larger.

Mercurius solubilis is the remedy when the head is so full that it seems it would burst, or as if it were laced tightly and drawn together with a band. Worse at night, tearing, burning, boring, stitching. [Hering.]

According to Black, Mercurius is useful in tearing, drawing headache seated in the pericranium and facial bones ; also in rheumatic headache with vertigo, and frontal headache with functional liver trouble.

Hemple recommends mercurius for syphilitic

headache with nocturnal paroxysms, which often increase to a frightful degree of intensity; hard, maddening bone-pains, as if the bones of the skull would be dashed to pieces; the patient is driven about the room by the violence of the pains.

In special aching of head and teeth, with affections of the mouth, jerking and beating pains in the forehead and temple, vertex; digging pains in the head, then tearing and stitching pain in the loose upper molars of either side, or extending from the side of the face back of the ear into the occiput, accompanied with dizziness, shivering, shuddering with flashes of heat, thirst, transient congestion, tendency to perspiration; great sensitiveness to each change of temperature; aggravation in the evening and at night; color of countenance sometimes yellow, with blue rings about the eyes. Rumbling in the bowels, pressure at the stomach, etc. Schelling commends mercurius.

Rademacher recommends it in the following conditions. In general he observed that the trouble was always periodic, sometimes quite as regular at first as an intermittent fever. The exacerbations more frequently occurred in the forenoon than in the afternoon; with a longer duration, the paroxysms became more and more irregular and ran into each other. The various forms are, the pain is located underneath the skull, and involves the region from the superciliary arch to the occiput. As the pain lessens, the external covering of the skull becomes painful.

II. The pain occupies the region of the superciliary arch on one side and the temporal bone, the eye on the same side being involved, with fiery redness and lachrymation.

III. Dull rheumatic pain above the orbits.

In one case the patient, who suffered from a dull, pressive pain, lost vision, with a loss of sensibility of the pupil. At the same time he appeared as if weak-minded, repeating the same thing again and again.

OPIUM.

Opium is suited to congestive, nervous, abdominal headaches; ailments from excessive mental emotion; joy, fright, anger, fear, or shame. The central difficulty is in the brain. Patient nervous and irritable, or inclined to stupor. Suited to old persons and recent cases. Sensitive pressure above right frontal eminence, with a sensation of heat while reading, giving place to a painful pinching sensation in the right temple, as if something pressed upon the part and then relaxed. Tendency of blood to the head, with constipation : violent tearing pains, or tensive pressure, through the whole brain, with beating or great heaviness in the head : unsteady look, thirst, dry mouth, sour eructations, desire to vomit.

Congestive headache with a sensation as if the

brain were constricted ; a stupefying pain, attended
with dizziness, tendency to sopor, obscuration of
sight. Sensation in the head like that following
sleep after excessive debauch. Brain oppressed,
extreme drowsiness. Dullness of the head ; has
no mental grasp for anything, and can not compre-
hend the sense of what he is reading. Very
sleepy, but can not go to sleep. Red, bloated,
swollen face ; eyes red, glistening, and prominent.
Renewal and aggravation of pains when becoming
heated.

The headache which is reached by opium often
depends upon cerebral disturbance or is closely
connected with disturbance in some other part of
the system, as in migraine.

Case.—A woman had, for many years, been subject to recurring headaches. Various means had
been used without obtaining relief. But the intensity of the pain and the increasing duration forced
the patient to seek help from new quarters with
each new attack. Symptoms : pressing, boring, or
stitching pain in the left supra-orbital region, increased by motion, noise and light. The pulse is
feverish. Chilliness, yawning, irritability of mind,
nausea. and vomiting at times precede and again
accompany the attacks. The urine is at first clear,
later it becomes darker and forms a deposit. The
temperature of the body varies at first, but soon increases moderately, and at last a gentle perspiration

makes its appearance, followed by an amelioration of all the symptoms. The attack is apt to come on at any time ; menstruation affects them only slightly. The weather seems to have more to do with them. During the intervals between the attacks, she is perfectly well. The patient receives opium during the attacks, and under its action they last but a few hours and come on less frequently. [Dr. Schmid.]

PLATINA.

Platina is adapted, according to Black, to headaches with leucorrhœa, of a nervous, tensive, cramping character, involving the forehead and orbital region. Headaches from anger or chagrin ; from uterine disease, with many hysterical symptoms. It is suited to females, dark hair, rigid fibre ; nervous spasmodic temperaments.

Neuralgic headaches, occurring in sensitive, fidgety, hysterical women ; with difficult or profuse menstruation ; cramp, pressing pain from within outward, with heat and redness of the face ; violent pressing in the forehead, roaring in the head ; worse from resting, in the room ; better in the fresh air and from motion. Violent crampy pains, especially over the root of the nose, with heat and redness of the face, restlessness, whining mood, roaring in the head as of water, with coldness of the ears, eyes, and one side of the face ; scintillations, illusions of sight, objects appearing smaller than they really are.

Cramp-like, drawing constrictions of the head, from time to time, especially about the forehead ; commencing slight, increasing until violent, and ending slight. Tensive, numb sensation in the zygomatic and mastoid processes, as if the head were screwed together. Crawling, like formication, on the right temple, afterward extending down along the lower jaw, with a feeling of coldness in it.

Vertigo, especially on sitting down, or going downstairs. Numb feeling in the brain. Physical symptoms disappear and mental symptoms appear, and *vice versa.* Pride, over-estimation of one's-self. Past events annoy.

The characteristic pain of platina is a cramp-like squeezing with numbness. Pains appear and disappear gradually.

RHUS TOXICODENDRON.

Rhus may sometimes prove useful in rheumatic headaches with tearing, stitching pains, extending to the ears, root of the nose, malar bones and jaws, with painfulness of the teeth and gums ; burning or beating pains ; fullness and oppressive heaviness of the head ; headache immediately after a meal ; desire to be quiet and lie down ; the pains are excited again by the least chagrin, or by walking in the open air ; wavering of the brain when stepping, and creeping in the head.

Head so heavy that it is necessary to hold it upright in order to relieve the weight pressing forward
into the forehead. Scalp sensitive, worse on side
not lain on, from growing warm in bed, from touch,
and combing the hair back. Headache in the occiput that disappears on bending the head backward.
Stupefying headache, with buzzing, formication and
throbbing ; face glistening and red ; restless moving
about. Heaviness and dullness of the head on
turning the eyes ; even the eyeballs hurt. On
shaking the head, sensation as if the brain were
loose and struck against the skull. The head is as
painful to touch as a boil.

Absence of mind, forgetful, difficult comprehen-
sion. General unhappiness of temper. Great apprehension at night, can not remain in bed. Melancholy, sad, depressed, discouraged. Impatient and
vexed at every trifle ; does not endure being talked
to. Pains are greatly aggravated by rest ; worse
after midnight and before storms ; relieved by
motion.

In 1839, when the type of all the prevalent diseases answered to rhus, there occurred cases of violent headache, one-sided, especially in the forehead,
temples, ears, and teeth, recurring periodically ;
drawing, tearing, as if a sharp knife were drawn
through the affected parts, or as if the ear, forehead
and eyes were cut off. The heartiest and strongest,
suffering from this pain, could not suppress loud

cries when thus taken. Arsenicum in some cases
gave prompt relief ; in several other rhus cured at
once. [Schelling.]

THUJA.

The sphere of thuja in the treatment of headaches is not extensive, being mainly limited to those
of sycotic or syphilitic origin, although it may prove
useful in some nervous or neuralgic pains. The
pains are of an intense stabbing character and almost unbearable ; worse from sitting, so that the
sufferer wants to lie down ; sitting up will sometimes cause unconsciousness from the violence of
the pain. The pains commence about the forehead
and eyes and extend backward. Intense pain, as
though a nail were being driven into the forehead
or vertex.

Tietzer uses this remedy in hemicrania when
there is reason to presume the existence of sycosis.
The pain by preference involves the left side, and
seems to extend in rays ; it is tearing in the forehead and face, as far as the zygoma, with a peculiar
sensation as if a convex button were being pressed
into the head, particularly in the neighborhood of
the sutures, or as if a needle were being crowded in
by jerks. If there is reason to presume sycotic taint
and the difficulty is of long standing ; when the
ache is always worse during rest, from warmth, especially in bed ; when it is reduced by looking up-

ward, with the head bent back ; when there are present rheumatic and gouty exacerbations, then thuja is the specific.

It is adapted to lymphatic temperaments, people with dark complexion, black hair, dry fibre, and not very fat. Extremely scrupulous about the least thing. Dissatisfied, quarrelsome, over-excited, angry at trifles. Can not think, talks slowly as if hunting for words ; uses wrong words.

Scalp sensitive to the touch or pressure of the pillow ; better if rubbed ; violent, tearing, burning, stitching pains, worse in warm bed. Vertigo ; with eyes shut, ceases on opening them ; when rising from sitting. Wants head and face warmly wrapped.

Case.—September 28, 1876, a gentleman called at my office whose wife suffered from intense headache. The pains are so severe, that she screams constantly and keeps the whole house awake. She nearly loses her consciousness and is unable to speak. Aggravation of the pains, and vomiting when rising up. Rest and horizontal position bring some relief, although the paroxysm is at its height about midnight. Another peculiarity is that the pain prevents the eyes from closing, and thus she had passed nearly two weeks with hardly any sleep, and the few snatches of sleep failed to refresh her. She felt weak and exhausted, especially as she is also troubled with habitual excessive menstruation, appearing too often and lasting too long. Although

only twenty-five years of age, she has already passed
through five puerperiæ. Thus there is a state of
anæmia with its consecutive painful nervous affections. The present attack of headache began with
great debility and lassitude, so that she had to go
to bed. During the first week the neuralgia was
bearable, but steadily increased during the second
week. The forehead, the region of the eyes and
ears, felt as if stabbed with knives, or, as she said,
as if knives went tearing around her brain. She
also complained of being chilly. She wanted to be
covered up, as her feet and knees felt cold. After
short intervals the pain always increased. There
was no thirst, but nausea and vomiting when rising
up, and frequent eructations. Palpitations were
frequently complained of. Prescribed thuja 100.
Cured. [Dr. Goullon, Jr.]

ZINCUM.

Zincum is indicated in several varieties of headache, one of which is an internal headache, mostly
semi-lateral, with stinging, tearing pains, worse from
drinking wine, or after dinner. It is also indicated
in obstinate pain in the head, which changes in intensity, at times severe and at times better, but persistent in duration. Pressure on top of head and
forehead, gradually increasing after dinner. Sensation of soreness on the vertex, as from ulceration ;
worse in the evening in bed, and at eating.

Chronic sick-headache ; great weakness of sight ; sticking in right eye. Chlorotic headache, especially in patients whose blood has been saturated with iron. Pressure on root of nose, as if it would be pressed into the head. Sharp pressure on small spot in forehead, evenings. Cramp-like tearing pain in right and left temple. Sufferings always relieved during flow of menses, return afterwards.

Dizziness, nausea, with vomiting of bile. Stupefaction and dullness of the intellect, especially afternoon and evening. Chronic cases of cerebral affections, great weakness of sight ; stitching pain in right eye ; paleness of face ; now and then vomiting.

Cerebral exhaustion, with mental and physical depression. Desponding, sad, low-spirited ; aversion to labor. Hair falls out on the vertex, causing complete baldness, with sensation of soreness in the scalp.

Case.—A woman, forty years old, strong, pale-looking, has suffered for two years with recurring attacks of headache, accompanied with such weakness of sight, that a heavy fog seems to rest upon her eyes and she cannot distinguish even large objects. There is a pressure from without towards within in the vertex and in the forehead, with a stupid feeling in the head, usually pallor of the countenance, loss of appetite, peevishness and irritability of temper. During the forenoon the pains are bearable, they grow harder in the afternoon, and in the evening, are very violent, at times accompanied with vomiting. Her bowels move every two or three days. Dimness of vision with the appearance of pains, increasing in ratio ; vision normal when she has no headache. The eyes present no objective symptoms. She has an attack every ten to fourteen days, coming on without warn- ing, lasting from two to three days and varying in severity during that time. Many remedies had been used unsuccessfully. Prescribed zincum met. 3, one powder morning and night. Improvement after the first dose, the attack continuing but one day. The headache did not again return. [Dr. Kafka.]

VERATRUM ALBUM.

Veratrum is a remedy of very great value in the treatment of headache, and although the range of its action is not extensive, the indications for its use are clearly defined. A characteristic symptom is the violent character of the pains, which drive to despair, great prostration, fainting, with cold sweat and great thirst, nausea, vomiting, pale face, stiff neck, profuse micturition ; sick headache in which diuresis forms a crisis.

Nervous headache at each menstrual molimen, neuralgia in the head, with indigestion, and sunken features, sensation in the brain, here and there, as though bruised, semi-lateral beating with pressure or constriction in the brain, with constriction of the throat, coldness on the top of the head, and cold sweat on the forehead, scalp very sensitive.

Vertigo : with cold sweat of the forehead ; with loss of vision, sudden faintings ; from abuse of tobacco, alcohol or opium.

Lobethal commends veratrum in the nervous headaches of young girls and hysterical women, with pale, earthy countenances, and nausea and vomiting. It is also adapted to young people and women of sanguine or nervo-sanguine temperament; also people who are habitually cold and deficient in vital reaction.

An excellent remedy, Black says, in pressing, beating, nervous headache, involving one side of the head, with stiffness in the nape of the neck, and a feeling as if the head would burst, accompanied with vomiting.

Hering regards it as useful when the scalp is painful upon combing and handling the hair; diarrhœa, pain so excessive that the patient raves; or the patient is remarkably weak, faint, worse from getting up and lying down; accompanied with cold sweat, coldness and thirst. Also when continued constipation gives rise to congestion to the head, with one-sided pain, pressing, beating, feeling as if the brain were bruised, with contractive feeling, or a constriction, extending into the throat, or with pain in the stomach, painfully stiff neck, copious evacuation of limpid urine, nausea, vomiting.

ACIDUM PICRICUM.

According to Hughes, picric acid is useful in severe headaches which begin in the occipital region, and thence extend forward and downward; the attacks are brought on by the slightest excitement or even the use of the brain.

Hempel believes it will prove valuable in brain fag or in utter exhaustion of the nervous system from long continued strain. Headache, with indifference and complete prostration.

Case.—Miss —— principal of a public school; tall, dark hair and eyes, consulted me in November last for a severe and terrible headache, which so completely incapacitated her for labor that she was compelled to relinquish the duties of her position. Her premonitory symptoms (which began in March, 1878) were, complete exhaustion, caused by her daily duties. She was so tired after her labors that she could scarcely reach home, although but a few blocks away. The headache began in the morning, on waking, increased as the day advanced and was always relieved by going to sleep at night. It affected chiefly the frontal region, extending gradually toward the vertex and involving the entire cerebrum, accompanied with constant vertigo, aggravated by motion and mental exertion—particu-

larly that connected with the duties of her school—
greatly increased by going upstairs; sometimes intense and throbbing, at others dull and pressing;
always better by keeping quiet. The pupils were
not dilated, although she had some pain in the eyeballs, which was increased by moving the eyes, but
not by reading. The face was never hot or flushed.
Accompanying the headache was a terrible sense
of prostration; "she was so tired." She felt better
in the open air than in the house, but was too
tired and exhausted to walk. She usually slept
well, but sleep did not always refresh her. No
night sweats. Appetite good; digestion and menstrual functions normal. Phosphorus 30 relieved
the headache somewhat; nux 30 and 200 relieved
the vertigo to some extent, but the feeling of utter
exhaustion continued. Picric acid 30 afforded
prompt relief and she resumed her duties again.
A slight return of the headache a month later
was promptly relieved by the 6th, and she now
feels better than she has felt in two years. [Dr. H.
C. Allen.]

PHOSPHORIC ACID.

Pressive pain with anæmic condition is the keynote of phosphoric acid. It is best suited to individuals of originally strong constitutions, which
have been weakened by loss of animal fluids, by
excesses, violent acute diseases, chagrin, or a long
succession of violent emotions.

Violent pressure in the forehead, in the morning on waking, with stupid feeling and inability to open the eyes. Dreadful pain on the top of the head, as though the brain was crushed, after long continued grief. The headache forces one to lie down and is insupportable ; aggravated from the least shaking or noise, especially music. Cerebral weakness from brain fag. School-girls' headaches. Pressure, as from weight in the head, from above downward.

Periosteal pains compel motion. Bones ache, feel as if scraped ; better on motion, when lying the pain shifts to the side on which he lies.

Dullness of the whole head, inability to think. Confusion and painful cloudiness of the head, especially on awaking. Chronic congestions to the head, caused by fright or grief. Hair turns gray early, especially after grief or sorrow.

ARGENTUM NITRICUM.

The characteristic symptom of this remedy is the sensation as if the painful part were enlarged ; if the pain is felt all over the head, it seems as if the head was enlarged ; if the pain is felt on one side of the head, the eye of that side appears enlarged. It is indicated in dull chronic headaches of literary and business men.

It is most suitable to nervous persons and to headaches from moral causes, characterized by dullness of the head, mental confusion, dizziness, tendency to fall, great weakness of the mind ; the headache is usually attended with chilliness and trembling of the body, intense nausea and vomiting. The head feels much enlarged ; time seems to pass too slowly. The patient is restless and in continual motion, or completely apathetic.

Hemicrania, pressive, screwing, throbbing pain in one frontal protuberance, temple, or into the bones of the face. Digging, cutting motion through the left hemisphere of the brain from occiput to frontal protuberance, recurs frequently, increases and decreases rapidly. Excessive congestion of blood to the head, with throbbing of the carotid arteries, obliging loosening of bands about the throat, accompanied with heaviness, stupefying dullness of the head, great melancholy, weakness of mind, inability to express himself rightly and coherently.

Case.—A cook, forty-four years old, has had repeated attacks of headache. The forehead, temple and vertex, at times the face and especially the superior maxillary, are never quite free from a pain, which is cutting and boring rather than tearing, and is increased at night in bed. Occasional remissions are followed with such violent hemicrania that she is forced to remain in bed and is nearly frantic. The pain then appears suddenly in the parietal bone, near the temple, is at first dull, increases rapidly and becomes throbbing and beating, terminating in nausea and vomiting. The attacks are preceded by pain in the hypochondrium ; the liver is hypertrophied ; slightly icteric. Indigestion, tasteless eructations, constipation, backache, heaviness through the hip, weariness and weakness of the legs. Prescribed argent. nitr. one drop each evening. Aggravation during the next two days, followed by marked improvement on the third day and prompt and early recovery. [Mueller.]

NAJA TRIPUDIANS.

According to Hughes, the headache of naja is a dull but severe pain in the temporo-frontal region, with much depression of spirits. Weight and pressure in the vertex, with cold feet and flushes of the face. It is also suitable for congestive headaches dependent upon organic disease of the heart. Periodic, neuralgic sick headache, very severe, in the left orbital region, extending back into the occiput ; the pain aching and throbbing about the orbit, and drawing in the occiput ; vomiting after the pain lasts for several hours.

Case.—Mrs. W., aged thirty-two years, of nervous temperament and feeble constitution, had been subject to attacks of neuralgia and sick headache for a number of years. She bore one child six years since, and following this period she suffered from some female weakness. From her youth she had been subject to periodical headaches, but subsequent to this they took on a very aggravated character, prostrating her greatly and confining her to the bed for days and weeks. The paroxysms were precipitated by irritability of the stomach, but the pain was usually severe for several hours before nausea and vomiting set in. The pain was very severe in the left orbital region and extended back into the occiput. About the orbit it was an aching pain ; after some hours it might be throbbing ; and thence it was drawing in its character to the back of the head. The vomiting varied from a watery and mucous to an acid or bilious vomiting. The bowels were usually regular and the menses normal. The patient had taken narcotics with palliative effects. Naja tripudians 6, in water, abated these distressing headaches and lengthened the intervals between the attacks. [Dr. D. A. Colton.]

CARBO VEGETABILIS.

Carbo veg. will be found useful in headaches depending upon the existence of a scorbutic or psoric diathesis in the organism. Aching or beating pains over the eyes or in the whole head, commencing at the nape of the neck ; the pains set in especially in the evening or after a meal, with tendency of blood to, and heat in the head. Hemicrania.

According to Hempel the headache to which carbo vegetabilis is homœopathic probably never

occurs without an accompaniment of such gastric
derangements as require the same drug for their
removal. In such headaches the patient will probably complain of an altered taste in the mouth,
bitter or sour; impaired appetite, tendency to
nausea and eructations, heartburn, fullness and
repletion of the stomach, bloating of the bowels,
costiveness.

Farrington recommends it for headache as the
result of dissipation; headache, particularly in the
morning when the patient awakes from sleep, having
spent the best part of the night carousing; dull
headache referred to the back part of the head, with
a great deal of confusion of mind. There is hum-
ing or buzzing in the head as though a hornet's nest
had taken its place there. The patient feels worse
in the warm room. The pains seem to go from the
occiput through the head and into and over the
eyes, giving a dull heavy aching in that region.
There are nausea and weakness referred to the
stomach, usually a burning sort of distress referred
to the epigastrium. He is unable to take any fat
food, whether meat, gravy, or fried food. He can
not drink milk because it produces flatulence. The
stomach feels heavy as if it were dragged down
after eating. The abdomen is distended with
flatulence. Both belching and borborygmi are
offensive. The wind belched has a rancid taste;
sometimes it has a putrid taste and a decidedly

offensive odor when passed from the bowels. He suffers from constipation with piles. The piles get worse every time he is on a spree. Sometimes they protrude and are bluish, they are so distended with blood. At other times he has morning diarrhœa with stool which is watery and thin and accompanied by a great deal of straining.

PHOSPHORUS.

Phosphorus is a valuable remedy in the treatment of headaches of nervous people or of persons who have exhausted their strength by sexual excesses or by too close application to study. Pulsating, throbbing pain with increased sensitive to odors. Headache, with drowsiness and inability to perform intellectual labor, relieved by quiet, and renewed by moving about, and accompanied by a sensation as if the brain were loose in the cranium. Sick headache, with pulsations and burning, mostly in the forehead, with nausea and vomiting, from morning until noon ; worse from music, while masticating, and in the warm room. Jerks in the sinciput as if pieces of lead were shaking in the brain. Hemicrania : The forehead or occiput is swollen ; touching the swollen part causes excruciating pain. Sensation of coldness in the cerebellum, with feeling of stiffness in the brain. Headache every other day. Chronic congestion of the brain. Softening of the brain, with persistent headache ; slow answering

questions ; vertigo ; feet drag ; formication and
numbness of the limbs.   Impending paralysis of the
brain and collapse ; burning pain in brain.   With
the headache, loss of senses as if he could not grasp
any thought.   Unable to collect the senses in the
morning ; on rising the head is dizzy, heavy, and
painful, as from lying with the neck too low.

Case.—A physician had an attack of rheumatic
arthritis of which he was cured after the lapse of
six weeks.   Since then he had been troubled with
attacks of headache which lasted sometimes a day
or more ; came on without any apparent cause, and
invaded at one time the forehead, at another the
occiput, etc.   The part where the pain was most
violent swelled, causing the most excruciating pain·
when touched ever so lightly.   The patient became
utterly incapacitated from performing the least
mental labor.   The left eye had become so weak
that he was no longer able to distinguish objects
clearly.   There was no visible change in the appearance of the eye.   In spite of the most careful and
abstemious mode of life, the patient had an attack
every few days which lasted from two to three days
without interruption.   The attacks were accompanied with an anxious choking and retching, lassitude in the limbs, an empty and confused feeling
in the head ; peevishness and depression of spirits ;
falling out of the hair and violent pains in the small
of the back after the attacks ; the pulse was ex-

tremely slow, about forty-five in the minute ; urine pale, watery, having a sickening, sweetish taste ; appetite undisturbed, except during the paroxysm, when he loathed all food ; felt very thirsty, and was tormented by so much uneasiness and anxiety that he was unable to contain himself. After having tried a number of remedies without benefit, he now took phosphorus, four grains dissolved in half an ounce of sulphuric ether, from twenty to twenty-five drops every two hours. Already after third dose he felt a pleasant warmth over the whole body, with frequent urination, cheerfulness, even mirthfulness ; the pulse was raised ; the horrid aching pain had become transformed into a dull pain ; the patient felt disposed to be quiet. He transpired over the whole body, perspired about the head, slept quietly all night, and felt very much refreshed the next morning ; pain and weariness had disappeared, the appetite had returned. He now took twenty-five drops every three hours. Next day the improvement continued. He felt cheerful, and a pleasant warmth over the whole body. The headache had entirely left him. Six weeks after he took a violent cold which commenced with a chill. The headache returned with great violence. He resumed the phosphorus in thirty-drop doses. After the first dose he felt a sensation of pleasant warmth over the body ; after the second dose he had a quiet and refreshing sleep for five hours. On waking a pro-

fuse perspiration had broken out; the headache
had left; he felt well and had regained his appetite.
On waking a profuse perspiration had broken out.
The urine, which had a fiery red color, deposited a
thick, white, slimy sediment two hours after standing. "This was," in the patient's own language,
" the last attack of this prostrating headache." [Dr.
Hempel.]

KALI BICHROMICUM.

Kali bichromicum is recommended by Farrington
in headache associated with the milder forms of
dyspepsia, the pain being usually supra-orbital. It
may be periodical in its return, but is particularly
excited by gastric irritation. Although neuralgic
in character, it is reflex from gastric irritation.
Another form associated with these gastric symptoms is headache, with blindness more or less
marked; objects become obscure and less distinct;
the headache then begins, it is violent, and is
attended by aversion to light and noise, and the
sight returns as the pain grows worse.

This remedy is suited to fat, flabby persons with
light hair and blue eyes. Persons subject to
catarrhs of the mucous surfaces. Headache which
involves the optic nerve. Comes in the morning,
the sight is blurred. As the pain increases in intensity the dimness of sight decreases. Comes and
goes with the sun and is accompanied with nausea.

Blindness followed by violent headache, must lie

down ; aversion to light and noise, sight returns with increasing headache. Violent shooting pains from root of the nose along left orbital arch to external angle of the eye, with dim sight, like a scale on the eye ; begins in the morning, increases till noon, and ceases toward evening. Burning headache with vertigo, during which all objects seemed to be enveloped in a yellow mist.

BOVISTA.

Bovista is indicated for menstrual headache lying deep in the brain, with feeling as though the head were enormously large or swollen. Morning headache, relieved by eating. Headache at 3 A.M., which gradually disappears after the breaking out of a profuse sweat. Headache on the right side in the morning, on the left side in the evening. On awaking the head aches as from too much sleep.

Absence of mind and difficulty of fixing the attention ; slowness of understanding ; moroseness ; ill humor. Headache is made worse by pressure and sitting up.

CACTUS GRANDIFLORUS.

Cactus grandiflorus is suitable for pressive headache occurring as a result of menorrhagia, or at the menopause, or for headache accompanied by or dependent upon cardiac disturbance.

Nervous sick headache, worse on the right side,

and from motion and strong light ; from wine, music, missing a meal, or dissipation.

Severe neuralgic headache from taking cold. Feeling as if the head were compressed in a vise, and that it would burst open from the severeness of the pain. Face bloated and red, with pulsating pain in the head. Severe pain in the right side of the head, worse from talking or a strong light. Pain commencing in the morning and growing worse as the day advances, until it becomes unbearable, with vomiting ; must lie perfectly quiet. Pressing pain in the head, as if a great weight lay on the vertex, increased by the sound of talking or any noise ; better from pressure.

GRAPHITES.

The curative effects of graphites are best shown in diseases due to the suppression of some eruption to which the drug was homœopathic, or because a scrofulous taint was at the root of the disease. In headache, the constitutional symptoms point to the remedy, such as weak digestion ; great hunger ; bloating of the abdomen ; constipation, large, difficult, knotty stools; eruptions oozing out a sticky fluid. The symptoms of the head are ; pain, as from bruises in the head, with general sick feeling in the evening. Violent headache, with eructations and nausea during menstruation. A pain as if the head were numb and pithy. Pain as if constricted, especially in the occiput, extending to the nape of the neck, which pains, on looking up, as if broken, at noon ; afterward the pain extends down the back and to the chest.

CASE.—Miss S., fifteen years old, large and well built, suffers every four weeks from a severe pain at the right temple. The pain is lancinating. Flick· ering before the eyes precedes and follows it. The headache may last one hour or a whole day. It is followed by general lassitude, deep slee}:, and the redness and heat of the head give place to chilliness. April 24.—Sepia 6, two drops in twelve teaspoonfuls of water, a teaspoonful morning and evening. After two weeks she reports that the headache has not returned, but flickering before the eyes still exists, with the general lassitude. She also complains especially of heaviness of the eyelids, which leads me to graphites ; and although she is fully developed, she has not yet menstruated, and a certain degree of hoarseness may be laid to the chronic hypertrophy of the tonsils. May 7.—Graphites, second trituration, two grains for six mornings. May 15.—Feels entirely well. Two years have since passed without any return of the headache or of the flickering. [Dr. H. Goullon, Jr.]

KALMIA LATIFOLIA.

Kalmia lat. is of value in cases of headache of a rheumatic or neuralgic character and origin ; with pain in various parts of the body; frequently accompanied with palpitation and irregular action of the heart, dyspnœa, vertigo and dimness of sight. Also for headache from exposure to the heat of the sun, pulsating, beating, pressing : worse on the right side ; if caused by exposure to the sun, the head feels better when the sun goes down. Headache internally, with the sensation, when turning, of something loose in the head diagonally across the top of it.

Case.—A servant girl had, for three months, been afflicted with a severe pressing headache which she named a " sun headache " because it increases with the ascending sun and decreases as the sun goes down. Having given her, at my office, a powder of kalmia, in the fifth trituration, she afterwards described the headache as suddenly going away as she waited in the car-office on her return home, and it never came back. [Dr. D. Thayer.]

Hepar Sulphuris.

Hepar sulphuris is suitable to scrofulous and debilitated subjects. The keynote is great sensitiveness to touch, draught of air or other impression on the senses. Rheumatic headache where the pressure of the hat is intolerable. Burrowing sharp headache, neuralgic in character. Headache from the abuse of mercury. Burrowing headache, as if a nail were being driven into the skull. Burrowing headache in frontal region as if an abscess were forming. The front of the head feels stupefied and heavy. Aching in the forehead like a boil, from midnight until morning. Violent headache, at night, as if the forehead would be torn out, with general heat, without thirst. Boring pain in root of nose, every morning from 7 to 12. Boring headache in right temple from without inward. Dull headache every morning in bed. Headache every morning, brought on by the least concussion. Sensitiveness of scalp to touch, with burning and aching, in the morning after rising. Great falling off of the hair.

Stomach inclined to be out of order, longing for sour or strong-tasting things. Sweats day and night without relief.

Natrum Carbonicum.

Natrum carbonicum is suited to leuco-phlegmatic constitutions with aversion to open air and dislike of exercise, physical or mental. The attacks occur during full moon, at certain hours, or on alternate days, and are characterized by a sensation of bursting, fullness in the head, stitching, tearing throbbing in the forehead, temples and vertex, with vertigo, melancholy, and great physical prostration. Pain from within outward.

Head feels too large and heavy, in the night and after dinner. Tension and obstruction in forehead, as if it would burst. Chronic headaches which are worse from exposure to the heat of the sun, with debility caused by the heat in summer. Chronic headache from sunstroke. Tearing pain in forehead, returning at certain hours of the day. Headache in temples from slightest mental exertion. Inability to think or perform any mental labor; the head feels stupefied if he tries to exert himself.

Avoids people ; seeks solitude ; full of apprehension. Restless, with anxiety ; aggravated by a thunderstorm or by music. Weight of eyelid, with dimness of vision, and floating spots.

PETROLEUM.

Petroleum is indicated more particularly in the headaches of persons affected with depraved mucous secretions, with tendency to lesions of the skin, and to the headaches of pregnant women who suffer much from nausea and gastric complaints.

Headache in the forehead, every mental exertion causes him to become quite stupid ; frontal headache, at times quite severe, but worse while the nausea remains. Pain from occiput over the head to the forehead and eyes, with transitory blindness, gets stiff, loses consciousness. Scalp very sore to the touch, followed by numbness, worse mornings and on becoming heated. Rapid appearance and disappearance of the symptoms.

PALLADIUM.

Palladium, according to Farrington, has a very

characteristic headache which makes the patient very irritable. It extends across the top of the head from ear to ear. The face is sallow, with blue rings around the eyes. There are also nausea, usually worse in the evening, and very acid eructations. The bowels are constipated, the stools often being whitish in color.

PARIS QUADRIFOLIA.

The headache peculiar to paris quadrifolia is of spinal origin, arising from the nape of the neck, and producing a feeling as though the head were immensely large. The scalp is sensitive to the touch, with sore pain at small spots on the forehead. The eyes feel heavy, as if they were projected, with sensation as if a thread were tightly drawn through the eyeballs and backward into the middle of the brain. Sense of weight and weariness in the nape of the neck and across the shoulders.

ALOES.

Aloes may prove useful in headaches which alternate with diarrhœa, or with pain in the back in hemorrhoidal patients. The pain is situated over the eyes, and is attended with a sensation as though a weight were pressing the eyelids down. Relief comes from partially closing the eyelids.

AMMONIUM CARBONICUM.

Ammonium carbonicum is especially applicable to fat, stout women, who lead sedentary lives and have headaches and various troubles in consequence. Congestive headache, with pulsating, beating, and pressing in forehead and vertex, as if it would burst at those points ; worse after eating ; while walking in open air ; better from pressure and in warm room. Headache with nausea, particularly early in the morning in bed, with pain in the stomach ; ill humor after dinner, lasting the whole day. Pain in the head resembling hammering or hacking with a flat instrument.

Vertigo, mostly in morning ; on moving head, sensation as if the brain moved to and fro toward the side on which he stoops ; scalp, even the brain, sensitive to touch ; sparks before the eyes at night; double vision ; buzzing in the ears with dullness of hearing.

PHELLANDRIUM AQUATICUM.

The characteristic symptoms of phellandrium are pain like a heavy weight, a stone, or a lump of lead, on the top of the head, with aching and burning in the temples and above the eyes ; pain in the eyes, with congestion of the conjunctiva ; watering of eyes ; intolerance of light and sound.

PLUMBUM.

The form of headache in which plumbum is curative is chronic dull headache, with depressed spirits and constipation.

Violent pains in the integuments of the skull,

from occiput to forehead.   Headache as if a ball were rising from the throat into the brain.

Great dryness of the hair ; it falls off, even in the beard.   General or partial paralysis, with wasting of the tissues.

NICKEL SULPHATE.

Periodical headaches returning every two weeks, and continuing two, three, or four days.   Acute pain at the root of the nose, extending to the vertex and through the temples.   Nausea without vomiting. Intense pain, confining to bed.

ANACARDIUM ORIENTALE.

The keynote for anacardium is the great mental irritability associated with the headache, which may be so great as to induce profanity.   Headache brought on by excessive mental labor, over-use of the brain.   Mental exertion brings on a tearing headache, the pain being situated mostly in the forehead.   In other cases the pain is similar to that of a plug in the head, or a constrictive sensation as though a band were tied about the head.   The pain increases hourly, momentarily relieved by hard pressure.

Great weakness of memory ;   can not remember anything ;   forgets everything immediately.   A slight offense causes excessive anger ;   breaking out in personal violence.   Feels as though he had two wills, one commanding to do what the other forbids.

Antimonium Crudum.

Headache in consequence of indigestion, or a chill, or recession of an eruption. Violent headache after bathing, with weakness of limbs and aversion to food. Splitting pain in the forehead, or else aching, boring, and spasmodic, or dull, tearing pains, especially in the forehead, temples, and vertex ; aggravations of the pains on going up-stairs ; amelioration in the open air. Chilly, aching sensation in the open air, Chilly, aching sensation in the limbs ; nausea, loathing, anorexia, risings and inclination to vomit.

Antimonium Tartaricum.

Headache from passive congestion of the brain ; sensation as if a band were tied around the forehead ; better from bathing the head ; from cool air and moving about. Throbbing pain, particularly on the right side of the head ; or drawing in the right temple, extending down and into the jawbone. Confusion of the head with feeling as if he ought to sleep. Mental excitement. Apprehensive and restless. Retching, and then vomiting, followed by great prostration.

Carbo Animalis.

The prominent symptom in carbo animalis is the feeling in the vertex as if pressed asunder ; must hold it together. Headache from menstrual suppression. It is adapted to complaints in nervous or

scrofulous constitutions. The attack is characterized by congestion of blood to the head, with heat
in the head ; vertigo ; confusion of the senses ;
changeable mood ; splitting, throbbing, or pressing
stitching pains, chiefly in the forehead and vertex.

Sensation as if something lay above the eyes, on
account of which she could not look up.

Great prostration and debility ; earthy-colored
face, with copper-colored spots on both the face
and body ; often troubled with severe headaches.

STICTA PULMONACEA.

Dull sensation in the head, with sharp, darting
pains through the vertex, side of the face, and lower
jaw ; dull heavy pressure in the forehead and at the
root of the nose. General confusion of ideas, inability to concentrate them ; but feels as if she must
talk, whether listened to or not. Catarrhal headache, before discharge sets in.

STANNUM.

The most valuable indication for stannum is the
gradual increase of the pain until it reaches its acme,
and then an equally gradual decline. It is most applicable to persons who are suffering from serious
nervous derangements. There is profound prostration of the nervous system ; patient must drop
down, but cannot very well get up. Pain every
morning over one or the other eye, mostly left,
gradually extending over the whole forehead, in-

creasing and decreasing gradually ; often with vomiting.

Acidum Hydrocyanicum.

Hydrocyanic acid may prove curative ·in congestive headache of a peculiar type. The headache is preceded by dizziness resembling intoxication. The patient complains of a dull, heavy pain in the fore part of the head, involving the sight, which is obscured, or it may first be felt in the occiput, whence it may extend to the forehead ; or it may be felt in various parts of the head. It is accompanied by extreme prostration, slowness of the pulse ; there may be oppression of breathing and feeling of embarrassment or weight in the region of the heart ; sensation of coldness with shivering.

Æthusa Cynapium.

Æthusa is sometimes useful in headaches of a periodical character, with sensation of tightness and pressure in the head and above the root of the nose. Restlessness and anxiety, driving into the open air, which relieves. Vomiting, with empty eructations and contractive, pressive pains at the epigastrium. Frequent and copious emissions of pale urine. Excessive sensitiveness of feeling.

Case.—A young lady has been subject for several months to headaches which come on periodically, appear suddenly, and last from three to four days. There is violent pressive pain in the

forehead, as is the head would burst ; eyes look as if
pressed outward, and there is marked pallor of the
countenance. Restlessness and anxiety, driving the
patient into the open air, where she finds much relief.
When the headache is most intense she vomits re-
peatedly, and experiences contractive, pressive pains
in the epigastrium, empty eructations, and, at times,
spasmodic hiccough. Frequent and copious emissions of clear, watery urine. The attack lasts several hours, and disappears gradually, usually followed by a watery evacuation from the bowels and
several hours sleep. After that she is well for the
next three or four days, with the exception of the
distress in the epigastrium. Aside from these paroxysms she is perfectly well. She received several
doses of æthusa after waking from the usual sleep,
which had followed an attack. After this she had
violent cramp-like pains at the stomach and vomiting of frothy matter which lasted nearly an hour.
She made a complete and permanent recovery.
[Dr. Hencke.]

BAPTISIA TINCTORIA.

The sphere of baptisia in the treatment of headache is confined to those which precede or accompany fevers of various types, with a sense of general
lassitude, indisposition, coated tongue, etc. If prescribed in the early stage of the disease it will not
only relieve the headache, but shorten the disease
as well.

CANNABIS SATIVA.

Cannabis sativa may be used in congestive head-ache, with aching pain in the frontal region, pale and sunken face, increased pulse, and general feeling of languor, sickness and difficulty of articulating. Continual headache on top of the head, as if a stone were pressing upon it. Sensation as if drops of water were falling on the head. Face pale, or the left cheek red, but not hot, and the right one pale. Pulse very weak, slow, frequently almost imperceptible. Violent beating of the heart, on moving the body, and on stooping, with warm sensation about the heart. Unendurable, fine stitching over the whole body, like from a thousand needle points, at night when sweating from warm covering, better when uncovering.

CANNABIS INDICA.

Sensation as of a heavy weight at the back of the head, from which pains shoot up the side of the head to temples and vertex ; worse at midday ; feels as if the head were opening and shutting at the top and the head was being lifted. Head feels very heavy, loses consciousness, and falls. On regaining consciousness violent shocks pass through the brain. Frequent involuntary shaking of the head. Every few minutes loses himself, and then rouses again to what is about him.

CAULOPHYLLUM THALICTROIDES.

Caulophyllum is indicated in rheumatic or neuralgic headaches dependent upon spinal or uterine diseases. Dullness of the head, with contracted feeling of the skin of the forehead. Headache with pressure behind the eyes and dimness of sight (with uterine disease). By spells, a very severe pain in the temples, as if both temples would be crushed together.

AGNUS CASTUS.

Agnus castus is best suited to headaches of persons with derangements of the womb, ovaries, testicles, or sexual organs in general ; the headaches of those given to sexual excesses, or subject to seminal emissions ; or those of unmarried persons suffering from nervous debility. Tearing pains, especially above the right eye and temple, as if one had received a blow upon the eye, attended with soreness to touch, increased by motion, aggravated in the evening, and lasting for several days. Headache in the upper part of the head, as from staying in a room filled with a thick and smoky atmosphere ; looking to one point relieves it.

ALUMINUM.

Alumina is suitable to persons of a scrofulous habit, who suffer from chronic diseases. It is indicated in chronic catarrhs of the head, and for headaches resulting from, or accompanying them. The head feels heavy, with paleness of the countenance and languor ; the vertex is painful to touch.

Stitches in the brain, sometimes with inclination to vomit.

CEDRON.

Cedron is principally adapted to persons of a voluptuous disposition, and of an excitable, nervous temperament ; more especially in females. There is a marked periodicity in the attacks, often with clock-like regularity. The pains are of a shooting character and located in the orbital region. Headache, especially deep in the orbits, obliging him to shut his eyes, and extending to the occiput.

EUPHRASIA.

Catarrhal headaches, accompanied by profuse, watery discharge from the nose and eyes. Violent headache as if bruised, or a sort of painful bewilderment of the head. Headache as if the skull would burst, with dazzling of the eyes from the light of the sun.

FORMICA.

Headache in the left fore part of the head and temples, back to the occiput, every day earlier, with a sore pain over the eye, beginning gradually, increasing, and with a cutting, extending into the ear. Headache in the posterior upper and inner part of the head, increased by drinking coffee and aggravated by washing in cold water.

DIGITALIS PURPUREA.

Digitalis acts best in headaches associated with

well-marked gastric disturbance or with profuse and frequent urination. The heart will also be found more or less involved, accompanied with an irregular, slow intermittent, or else small, unequal, and very rapid pulse, excited by the least movement.

ÆSCULUS HIPPOCASTANUM.

Æsculus hip. is suitable for headaches in persons of hemorrhoidal tendencies, and who suffer with gastric, bilious, or catarrhal affections. The headache is dull rather than acute, pains here and there, chiefly in the right temple and occiput, followed by dull stitches in the forehead and left temple. Sensation of fullness and pressure more than acute pain. Feeling as if a board were on the head ; confusion of the head and giddiness.

ARANEA DIADEMA.

The characteristic symptom of aranea is the aggravation from dampness ; either a damp day or a damp place will induce or aggravate an attack. It is adapted to persons of a hydrogenoid constitution, when the attacks come at regular hours ; glimmering before the eyes; dizziness which causes the patient to lie down ; on rising, a feeling as if the head and hands were bloated and swollen. Headache and confusion of the head are relieved by smoking, but return again. Sudden, violent pains in the teeth of the whole upper jaw, and also the lower one, at night, immediately after lying down.

ASCLEPIAS SYRIACA.

Asclepias may be curative in congestive headache from suppression of sweat or urine, and fever. Nervous headaches, which are attended by dry skin and scanty urine, cool skin and feeble pulse, and followed by sweating or profuse urination. It may also prove useful in rheumatic headache. Headache from suppressed perspiration, or from the retention of effete matters in the system. A feeling as if some sharp instrument was thrust through from one temple to the other, with feeble pulse and cool skin.

CORNUS CIRCINATA.

The characteristic symptom of cornus cir., is sense of fullness in the head, relieved by a copious stool. The pain is dull and heavy through the whole head, with drowsiness ; it is increased by walking, stooping, or shaking the head. Dull, heavy sensations in the head ; with shooting, aching, and throbbing pain. Sense of fullness and pressure in the head, preventing sound sleep.

EUPATORIUM PERFOLIATUM.

The headache of eupatorium is of bilious origin, and characterized by its periodicity. There is pain in the occiput after lying, with sensation of great weight in the part, requiring the hands to lift it ; headache better in the house, aggravated when first going into the open air ; relieved by conversation ;

throbbing headache. Darting pain through the temples, with sensation of blood rushing across the head ; soreness and beating in the back part of the head.

HELONIAS DIOICA.

The characteristic indication for helonias is headache relieved by motion or mental exertion, but which returns as soon as the patient is quiet. Pain in the forehead, as if a band about an inch wide were drawn across the temples ; feeling of fullness in the head, with vertigo ; pain in the occiput, with pulsative pain in the vertex ; increased by stooping ; attended with vertigo ; great activity of the salivary glands.

LACHNANTHES TINCTORIA.

Lachnanthes may be useful in headaches of a congestive and neuralgic type, especially when the eyes are much affected. The characteristic symptoms are · sensation as if the head were enlarged, and split open with a wedge from the outside to within ; body is icy cold, cannot get it warm ; the head burns like fire, with much thirst. Sensation as if the vertex were enlarged and were driven upward. Feeling as if the hair were standing on end, with soreness of the scalp. Tearing from the right side of the forehead into the cheek.

LEPTANDRA VIRGINICA.

Leptandra is indicated in bilious headaches, with

constant, dull, frontal headache, with dizziness while walking, accompanied by constipation or diarrhœa, furred tongue, bitter taste, indigestion, yellow urine, languor and depression of spirits.

LOBELIA INFLATA.

The headache of lobelia is always accompanied by irritation of the pneumogastric nerve, and usually also by great muscular prostration and mental despondency.   There is dull, heavy pain around the forehead, from one temple to the other, on a line immediately above the eyebrows.   Pain through the head in sudden shocks.   Continual periodical headache in the afternoon, increasing until midnight, every third attack being alternately more or less violent.   Sick headache, with vertigo ; dull headache, violent nausea, vomiting and great prostration.

PHYTOLACCA DECANDRA.

Phytolacca will be found of especial value in persons of a syphilitic habit, with much soreness deep in the brain, of irritable and depressed mood ; suffering from vertigo and dimness of sight ; aggravated by damp weather.   Dull, heavy pressing pain through the whole head.   Sick headache, worse in the forehead, with backache and bearing down ; comes every week.

SELENIUM.

Selenium is indicated in headaches of nervous origin ; pain of a stinging character, usually over the left eye ; worse from the heat of the sun.   The pain returns, periodically, every afternoon, and is increased by any strong odors.   It is worse from drinking tea, lemonade.   The headache is associated with profound melancholy and attended with a profuse flow of limpid urine.   There is also nausea, heavy coated tongue, and bilious vomiting.   The patient has a desire for brandy, which sometimes relieves the headache as well as the gastric symptoms.

THERIDION CURASSAVICUM.

The keynote of theridion is hypersensitiveness. Every penetrating sound and reverberation penetrates through the whole body.   The head aches worse if any one walks across the floor.   There is sharp neuralgic pain over the left eye, with weakness, trembling, coldness and anxiety.   Very severe headache, with nausea and vomiting, like sea-sickness, and with shaking chills.   Headache on beginning to move.   Headache behind the eyes, with luminous vibrations before the eyes.   Vertigo with blindness, from the pain in the eyes.   Head feels thick ; thinks that it belongs to another ; that she can lift it off.

CAPSICUM ANNUM.

In catarrhal, gastric, rheumatic and nervous headaches capsicum may prove a useful remedy.   It is adapted to phlegmatic constitutions with hemorrhoidal tendency ; or to indolent persons of suspicious dispositions, or to headstrong, clumsy people, afraid of exercise and the open air. The headache is characterized by a sensation of congestion or fullness of the head ; throbbing, pressing or tearing pains ; feeling as if the skull would burst, especially when moving ; pain mostly in forehead and temples ; darting pains in head ; skull feels bruised. Confusion of the head with obscuration ; dizziness and dullness of head with febrile chill and coldness ; nausea and vomiting. Headache on coughing, as if the skull would burst.

LYCOPODIUM CLAVATUM.

Lycopodium may be found useful in headaches based upon gastric or bilious disturbances, or in headaches caused by chagrin ; the characteristic indications are : disposition to faint and great restlessness. The pain increases or comes on every afternoon, from 4 to 8 P.M. The head aches as if the bones of the skull were being driven asunder, and as if the brain were vacillating ; or dull pain in the forehead as if the head were being compressed. The pains are worse from the warmth of the bed, or warmth generally, and from mental exertion. Lycopodium affects more the right side.

MELILOTUS.

Violent cerebral congestion with headache, which drives the patient almost frantic. Feeling as though the brain would burst through the forehead. Intense throbbing pain.

OLEANDRA.

The characteristic symptom of oleander is headache relieved by looking sideways or cross-eyed. Headache with vertigo ; pressive pain from within outward ; biting itching on the scalp, as from vermin, worse on the back part of the head and behind the ears, better when first scratching it ; followed by burning and soreness, which gives place to biting itching ; worse in the evening when undressing. Stupefying pressure at the malar bone, penetrating into the brain, and extending to the root of the nose ; stupefying, troublesome feeling of tension. Violent pressing pain in the temples when swallowing ; great general weakness ; pale face with blue rings around the eyes, feeling of tightness and suffocation at the throat.

Case.—Miss A., aged twenty-one years, complained of headache, which was improved by looking cross-eyed. There were no other symptoms that would guide me. I gave two doses of oleander 200. The result was a cure. [Dr. E. A. Farrington.]

PSORINUM.

When the well chosen remedy does not act, psorinum may prove useful, though not often indicated. It is adapted to scrofulous, nervous, restless people, or to pale, sickly delicate children. The head aches as if the brain had not room enough, when rising in the morning, better after washing and eating. Chronic headache, aggravated at every change in the night while sleeping ; is awakened by the pain. Wakens at night as from a heavy blow on the forehead. Pain like hammers striking the head from within outwards. Mental labor causes fullness of head ; intense headache ; throbbing in brain ; pain in left temple. Is always very hungry during headaches. Headaches from repelled eruption ; the pain is preceded by specters, by dimness of sight, and spots before the eyes.

RANUNCULUS BULBOSA.

Ranunculus bulb. has headache on the vertex as if pressed asunder, with pressure on the eyeballs and sleepiness ; worse in the evening, or from changes of temperature. The headache is caused by changes from a warm to a cold, or from a cold to a warmer temperature. The pains are also excited or aggravated by contact, motion, stretching, or changing the position of the body.

LAC DEFLORATUM.

Lac defloratum, Farrington says, has cured intense headache located principally in the fore part of the head. The pains are of a throbbing character and are associated with nausea, vomiting and the most obstinate constipation.

Lithium Carbonicum.

Lithium carbonicum has not an extensive range of action. There is confusion of the head, headache on the vertex and on the temples, worse on awaking ; the eyes pain as if sore, with difficulty of keeping the lids open ; worse from morning to noon. Pressure in the left temple with weight on vertex, relieved while eating, worse after eating. Pain and heaviness over the brow ; aggravated toward evening, continued until falling asleep. Sunlight blinds ; inability to see the right half of objects. The headache and soreness of the eyes follow on suppression of the menses.

Magnesia Muriatica.

Magnesia muriatica is indicated in congestive headaches with sensation of boiling water in the brain. Pain in the temples better from pressure with the hands and from wrapping the head up warmly. It is suitable in hysteric and spasmodic headache, especially in women with uterine affections. Sensitive to cold air. Attacks accompanied with stupefaction and dullness of head, with compressive pain and numbness in and about the head ; general relief from strong pressure. Headaches, especially chronic and in persons of constipated habits. Very severe pain in both temples, with dizziness, and loss of consciousness.

Case.—A clerk, forty-eight years old, of a yellowish-brown complexion, choleric, suffered for several years from severe headache. The pains are usually located in the forehead and around the eyes ; it seems as if the head would burst ; he is obliged to lie down ; the headache is aggravated by motion and in the open air ; warmth does not relieve, but strong pressure does. At times these attacks continue for several weeks, coming on every day, and they return about every six weeks. He also complains of loss of appetite, bitter taste, eructations, water brash ; the epigastric and the hepatic region is sensitive to pressure ; the liver is hard and enlarged ; the patient can not lie on the right side ; the bowels moved only with the aid of an enema, when he passed some small yellowish-gray balls ; urine of a deep yellow color, with considerable mucus ; the tongue is thickly coated, with clean tip and edges ; great thirst ; no fever. Magnesia mur. cured the case in three weeks. Three years have passed and he has had no relapse. [Dr. Stens, Jr.]

Acidum Carbolicum.

Carbolic acid has been used with good effect in monthly headache, before, during or soon after the menses ; the seat of the pain is over the right eye ; the brain is compressed as if in a tight bandage. Sickness of the stomach at times during the attack.

Causticum.

Headache, with blindness, which does not diminish as the headache increases. Rheumatic or arthritic diathesis. Frequently suitable in cases with a tendency to scrofula or having a weakened and emaciated appearance. Sensation as of an empty space between forehead and brain, or as if something was forced in between frontal bone and the cerebrum. Pain at a small spot on the vertex, as if bruised, only on touch. Sudden and frequent loss of sight, with a sensation of a film before the eyes. Dull pain in the ears, with constant whizzing, as though steam escaped.

CYCLAMEN.

The marked symptom in cyclamen is sudden vanishing of the sight, with semi-lateral headache of the left temple, with pale face, nausea referred to the throat, and weak digestion. It is suited to blonde leuco-phlegmatic subjects with chlorotic condition ; disinclination for any kind of labor ; easily fatigued ; special senses enfeebled or with functions suspended.

Violent headache with flickering before the eyes, on rising in the morning. Slight pressure in the vertex as if the brain was bound with a cloth, which would deprive him of his senses. Fine, sharp, itching, stinging in the scalp, which always reappears in another place on scratching, worse evenings and at rest ; relieved by motion. Feels as if the brain was in motion when he leans against anything. With headache one always sees countless stars.

IPECACUANHA.

Ipecacuanha is sometimes indicated in headaches of rheumatic origin. The characteristic sensation is a pain as if the head or bones of the head were bruised or crushed, this feeling seeming to go down into the root of the tongue. This headache is accompanied by nausea and vomiting. It is also useful in sick-headache ; the pain is confined to one side of the head, with deathly nausea. The face is pale, with blue rings around the eyes.

JUGLANS CATHARTICA.

Recommended by Farrington for occipital headache with pains of a sharp character.

LAUROCERASUS.

Chronic headaches, or the headache from nervous prostration, are amenable to laurocerasus. The pains are dragging, boring, always accompanied with drowsiness. It is also curative in headaches of hepatic origin, with pain and swelling in the hypochondrium and yellow spots on the face. Sensation of icy coldness on vertex, as from cold wind, then in the forehead and nape of the neck, extending to the small of the back, after which all the pains in the head disappear. Pulse small, slow and contracted.

NUX MOSCHATA.

Nux moschata is especially applicable to hysterical and gastric headaches of people of a highly sensitive nervous organization ; after suppression of some eruption. Headache, worse from getting wet, change of weather, riding in a carriage, after eating or wine, from suppressed eruptions, before menses, during pregnancy. Throbbing, pressing pain, confined to small spots. Region of temples very sensitive to pressure. Pain, especially in the temples ; on shaking the head there is a feeling of looseness, as if the brain beat against the skull. The head inclines to drop forward, the chin resting on the breast, and it could only be raised with much effort, after which it again drops forward. Vertigo, as if drunk, staggering ; weak limbs, numb ; feels as if floating in the air. The head feels full, as if expanding. All ailments are accompanied by sleepiness, and inclination to faint.

Case.—A strong, thick-set man has suffered from a headache for a period of four weeks ; it is characterized by soreness, pressure, sensation as if the head would burst ; the seat of the pain is not deep in the brain, but rather immediately under the skull ; when perfectly quiet it becomes eased and finally disappears ; returns, however, upon every motion, stooping or shaking the head. The memory is impaired. Cured by nux moschata. [Dr. Montgomery.]

HELLEBORUS NIGER.

Pressing headache from outward inwardly, with stupefaction and heaviness of the head, worse from moving the head, from exertion, better in the open air, and from distraction of the mind. Stupefying headache, with coryza (4 to 8 P.M.); worse from stooping, better from rest and in the open air. Shocks pass through the brain like electricity.

HYDRASTIS CANADENSIS.

Suitable to headaches of a catarrhal origin. Headache accompanying dyspepsia occasioned by derangement of the mucous coat of the stomach. Useful in headaches occurring as a symptom of uterine disorder. Especially adapted to weakened, debilitated subjects, with mucous discharges. Headache of a nervous, gastric character, almost constant. Myalgic headache, in the integument of the scalp, and muscles of the neck. Constant, dull headache, with pain in the hypogastrium and small of the back, of a dull, aching character.

SABADILLA.

Sabadilla may prove a valuable remedy in some forms of nervous and congestive headaches. Vertigo with constant nausea ; relieved by lying down. Pressure and heaviness in the head ; can scarcely raise it. A dull, oppressive ache in the anterior portion of the head, abating by pressing the flat hand against the forehead ; feeling of increased warmth, which is followed by a continued coldness in the hairy scalp ; even the hair feels cold to the hand, as if cold water had been poured over the head. Stupefying, aching pain in the forehead, causing a feeling of giddiness, and making him stagger from side to side as if intoxicated. Headache as if a thread had been drawn through the brain, over the temples, from the forehead toward the occiput, leaving a burning sensation behind. Sensation as if the head were in a vise. Headache after every walk; on re-entering the room, a wrenching, screwing pain commences in the right side of the head, seizing upon both temples, where it is felt very keenly, whence it spreads over the whole head after retiring ; it returns every day.

INDIGO.

The pains are characterized by great intensity, are worst during rest and when sitting, and can frequently be entirely suppressed by rubbing and pressure, or by motion, or alleviated so as to reappear with less intensity. The majority of the pains come on, or are aggravated, in the afternoon and evening.

LILIUM TIGRINUM.

Attacks occur in those affected with uterine troubles, menstrual irregularities and irritable condition of the heart. Pains in forehead and temples, with vertigo, depression of spirits, bearing down pains in region of womb and anus, strangury and ineffectual urging to stool. Pain through the whole head, with sensation of fullness ; feeling as if blood would is-

sue through the nose when blowing it ; amelioration
from supporting the head with the hands, and at
sunset ; aggravation when walking in the open air.
Drawing hot pain through the head and eyes,
ameliorated by frequent sneezing at 10 P.M.
Stinging, burning pain in the forehead, with sensation as if a rubber band were stretched over it,
with confusion of mind ; relieved by effort of will.
Depression of spirits ; constant inclination to weep,
with fearfulness and apprehension of suffering from
some terrible internal disease already seated. Firm
conviction that the disease is incurable. Constant
hurried feeling, as of imperative duties, and utter
inability to perform them. Symptoms return again
and again after having disappeared, but in a dimin-
ished degree.   Social and moral conditions pro-
foundly affected, generally changed to the opposites.
Symptoms aggravated by repose and thinking about
them ; are ameliorated when busy and in the open
air.

MERCURIUS IOD-FLAVUS.

The headache of this remedy is felt more while
at rest, being relieved when the mind or body is
actively engaged. The pain is principally on the
right side or on the top of the head. Sensation as
if the skull were cracking.

TARANTULA.

Sharp, nail-like pains in the head. Intense
headache, as if thousands of needles were pricking

into the brain ; better from rubbing head against the pillow. Headache, as if a large quantity of cold water was poured on the head ; relieved by pressure.

CHIONANTHUS VIRGINICA.

In cases of habitual sick-headache, 5 gtts. of the 2x dil. three times a day for a week, then twice a day for a week, then once a day for a week—after which only when symptoms of the attack show themselves. Severe headache chiefly in the forehead, just over the eyes—especially the left eye. Eyeballs painful, with sore, bruised feeling (cutting, twisting pain in abdomen relieved by lying upon it); drawing and pressing at the root of nose. Tongue heavily coated of a greenish-yellow color. Great nausea and retching, with desire for stool, vomiting of dark-green, ropy bile, very bitter ; cold perspiration, and prostration.

Dr. Gross calls attention to the drug as having special action upon the hair, and being capable of producing the typical sick headaches and jaundice. [Dr. M. H. Van Tine.]

EPIPHEGUS VIRGINIANA.

Is adapted to that form of sick-headache which accompanies nervous exhaustion. Every unusual effort—any excitement, pleasurable or otherwise, a day's shopping, loss of sleep, fatigue or anxiety—in short, any demand which exhausts the energies, brings on an attack. This is often acompanied by nausea, and vomiting which does not relieve. The pain is through the temples and all over the head. It is said to cause severe pain in forehead with fullness, dull and heavy, feeling of tightness becoming more severe. Blurred vision, eyes smart ; confusion of mind ; makes wrong letters, uses wrong words. Nausea, *with constant desire to spit ;* viscid saliva. Feels worse on rising and moving about, better from rest and when reclining ; sleep invariably brings relief. For bilious and menstrual headaches, seldom gives relief ; when it does there is probably some element of nervous exhaustion present.

It is given 3 D. gtt. 5 in water twice daily between the attacks. During the paroxysms every twenty or thirty minutes. [Dr. M. H. Van Tine.]

USNEA BARBATA.

Resembles glonoine in the peculiar congestions which it produces. Like that drug it is useful in cases of sunstroke. There is throbbing of the carotids, with the red flushed face, and staring blood-shot eyes, and pressing (usually) frontal headache. In a case of this kind, the writer found it serviceable where other remedies had failed, but on a return of the attack, a day or two after, it was apparently quite inert, while Bell. 200 gave immediate relief, and continued to do so whenever there was any recurrence of the pain, which was very severe at intervals for a day or two. [Dr. M. H. Van Tine.]

Salicylate of Sodium.

Salicylate of sodium is said to remove the gastro-intestinal disturbances along with the headache. In its action it resembles calomel, and, like that drug, it frees the secretions of the mouth, and at the same time relaxes the bowels. Dose, 2 or 3 grs. every quarter or half-hour until three or four doses are given. Begin when the pain first comes on.

ANÆMIC.—Agn., ars., bry., china, cocc., dig., epiph,, ferr., gels., lac deflor., nat. mur., nux v., pall., puls., phos., sulph.

CATARRHAL.—Acon., æscul., hip., am. mur., ars., bell., bry., carbo v., caul., cham., china, cimici., dulc., euph., gels., hepar, hyd., ign., kali bi., lach., lyc., merc., nat. mur., nux v., puls., ran., sang., sulph.

HYPERÆMIC.—Acon., agar., am. carb., arn., arg., asclep., aur., bell., bry., calc. carb., camph., cann., caps., carbo v., caust., cham., china, cimici., cocc., dig., dulc., feri., gels., glon., hel., hepar, hyd. ac., ign., lach., lil., lyc., melil., merc., naja., nux v., opi., phos., phos. ac , pic. ac., puls., rhus tox., sang., sep., sil., spig. sulp., ther., usnea.

NERVOUS.—Acon., agar., agnus, anac., arg. nit., arn., ars., asar., asclep., aur., bell., bry., cact., calc., carb., camph., china, cimici., cann., caul., caust., cedron., cham., cocc., coff., coloc., croc., epiph., gels., glon., graph., hyd., ign., ipec., iris, nat. mur., nux. v., opi., petr., phos., phos. ac., plat., puls., rhus tox., sang., sep., sil., spig., sulph., tarant., ther., thuja., verat.

NEURALGIC.—Acon., am. mur., arn., ars., bell., bry., cact., calc. carb., caps., caul,, caust., cham., china, cimici., coff., coloc., gels., hepar., ign., kalm., lach., lyc., merc., naja, nux v., phos., plat., puls., sang., sil., spig., stann., sulph., thuja., verat.

REFLEX.—Uterine—Agnus, ars., bell., bov., bry., cact., calc. carb., caul., cham., cimici., cocc., coff., col., croc., gels., hel., ign., lach., lil., lyc., nat. mur., nux mos., nux v., plat., puls., sang., sep., sil.. sulph., thuja., verat., zinc.

DYSPEPTIC.—Acon., æsculus., am. carb., anac., ant. crud., arn., ars., asar., bell., bry., calc. carb., caps., caust., caul., carbo veg., cham., china, cimici., cocc., coloc., gels., glon., hyd., ign., ipec., iris, kali. bi., lach., lyc., naja, nux v., opi., paris, phos., plat., puls., sang., sep., sil., sulph., tarant., verat.

RHEUMATIC.—Acon., am. mur., asclep., bell., bry., caul., caust., cham., china, cimici., coloc., dulc., ign., kali bi., kalm., lach., lyc., magn. mur., merc., nux v., phos., pul., rhus tox., sep., sil., spig., sulph.

TOXIC.—Asclep., ant. crud., arn., ars., aur., bell., calc. carb., caps., carbo v., caust., cham., china, cimici., cocc., dig., gels., hep., ign., ipec., lach., lyc., nat. mur., nux v., rhus, puls., sel., verat., zinc.

CAUSES.

Air, cold—Ferr., nat. mur.

Air, exposure of head to cold, damp currents of—Dulc., cimici.

Air, exposure of head to cold, dry currents of—Acon.

Air, a current of—Acon., ars., bell., china, coloc., nux v.

Air, exercise in the open—Am. carb., calc. carb., hep., pet.

Anger, contradiction—Acon., anac., cham., coff., coloc., lyc., nat. mur., nux v., opi., petr., phos., phos. ac., rhus tox.

Animal fluids, loss of—China, nat. mur., phos. ac.

Bathing—Ant. crud., calc. carb., puls., rhus.

Beer drinking—Cocc., rhus.

Carriage, motion of—Cocc., graph.

Chamomile tea, abuse of—puls.

Chill—Acon., ant. crud., bell., bry., calc. carb., cham., coff., hep., ign., lyc., merc., nux v., puls.

Coffee, abuse of—Bell., caust., cham., cocc., hep., ign., lyc., merc., nux v., puls.

Cold and dampness—Bry., dulc., rhus, spig.

Cold—Acon., ant. crud., bell., bry., cham., china, coloc., dulc., nux. v., phos., puls.

Cold drinks—Acon., ars., bell., puls.

Cold, sudden change to—Dulc.,

Concussion—Arn., bell., cocc., hep., phos.

Copper, abuse of—Hepar.

Cutting hair—Bell., phos.

Drugs, abuse of—Nux v.

Emotional excitement—Epiph., gels., ign., phos. ac.

Fatigue—Epiph.

Fright—Acon., opi., phos. ac.

Grief—Ign., phos., phos. ac.

Heat—Acon., am. carb., arn., bell., bry., caps., carbo v., ign., ipec., sil.

Heat of sun—Glon., lach., nat carb., nat. mur., nux v.

High living—Nux v., puls.

Injuries, mechanical—Arn., merc., petr., rhus.

Intellectual labor—Anac., arn., asar., arg. nit., aur., bell., calc. carb., china, cimici., coff., coloc., dig., graph., ign., lach., lyc., nat. carb., nat.

mur., nux v., nux mos., petr., phos., puls., sep., sulph.

Iron, abuse of—Puls., zinc.

Joy—Coff., epiph., opi.

Lemonade—Sel.

Malaria—Ant. crud., arn., ars., calc. carb., caps., carbo. v., china, dig., gels., ign., ipec., lach., lyc., merc., nat. mur., nux. v., puls., rhus., sulph.

Mercury, abuse of—Aur., carbo v., china, hepar, puls., sulph.

Metallic substances—Sulph.

Onanism—China, gels., nux v., phos. ac., sulph.

Opium, abuse of—Bell.

Overheating—Bell., bry., carbo v., sil.

Quinine, abuse of—Arn., ars., carbo v., gels., ign., ipec., lach., nat. mur., nux v., puls., sulph.

Sexual excess—Agn., china, gels., nux v., phos. ac., sulph., thuja.

Shame—Opium.

Sleep, loss of—Epiph.

Spiritous liquors, abuse of—Ant. crud., ars., bell., bry., calc. carb., carbo v., china, cimici., coff., ipec., nux v., phos., rhus, sulph.

Strain, lifting—Arn., bry., calc. carb., phos. ac., rhus., sil.

Sulphur, abuse of—Puls.

Suppression of eruption—Acon., asclep.

Syphilis, effects of—Aur., hepar, merc., thuja.

Tea, abuse of—Sel.

Tobacco, abuse of—Acon., ant. crud., cocc., ign.

Watching, prolonged—Bry., colc., china, cocc., nux v., puls., sulph.

Weather, variable—Ars., bry., carbo v., nux v., rhus.

CONDITIONS.

Air, in open—Bell., calc. carb., china, ferr., kalm., lach., nux v., spig., sulph.

Air, after exercise in open—Am. carb., calc. c., hep., petr.

Air, during exercise in open—Ant. crud., kalm., nux v., puls., rhus, spig., sulph.

Awakened by the pain—Psor.

Awakening at night—Chel.

Awaking, on—Bov., naja, phos., phos. ac.

Bending body backward, when—Colch.

Bed, after going to—Ars., lyc., magn. mur., puls. sep., sulph., zinc.

Bed, obliges one to leave—Coloc., ferr.

Blindness, before—Gels., kali bi.

Blindness, with—Caust., nat. mur., iris v., psor.

Blindness, after—Sil.

Blowing the nose—Sulph.

Climaxis, during—Carbo v., croc., lach.

Climaxis, during, when menses used to appear—Croc.

Cold in head, preceding—Lach.

Commencing slight, increasing till violent, ending slight—Plat.

Coughing, after—Ipec., stan.

Coughing, when—Anac., caps., carbo v., kalm., spig., sulph.

Dark, must lie in, and perfectly still—Sang.

Despair, violence of pain drives to—Verat.

Drinking, after—Acon.

Drinking and eating, after—Bell., colc.

Eating, after—Am. carb., arn., ars., calc. carb. an., carbo v., cham., graph., kalm., lach., lyc., nux m., nux v., phos., puls., rhus, sulph., zinc.

Eating, when—Graph.

Evening, in the—Am. carb., anac., carbo v., cham., croc., euphr., fer., hep., lach., lob., lyc., petr., phos., rhus, sep., sulph., ther., thuja, zinc.

Evening, in the, in bed—Ars., lyc., puls., sep., sulph., zinc.

Exertion, physical—Anac., calc. carb.

Eyelids, upper eyelids so heavy the eyes almost close—Gels.

Eyes, on opening the—Bry., china.

Eyes, when moving and turning the—Bell., bry., chel., dig., hep., nux v., opi., puls., rhus.

Eyes, obliges him to shut the—Cedron, cocc.

Eyes, always beginning with a blur before the—Iris.

Eyes, prevents from opening—Bell., sep.

Eyes drawn shut by the severity of parietal head-ache—Bell.

Eyes must be kept closed and the head constant-ly in motion—Agar.

Eyelids can only be raised with exertion and pain—Nat. mur.

Faints in consequence of pains—Gels.

Frowning, when—Nat. mur.

Gaslight, while working under—Glon.

Going up, on—Calc. carb., sulph.

Going up a height, on—Calc. carb.

Going up stairs, when—Ant. crud., arn., bell., lob., paris, phos. ac.

Gradually increasing, then gradually decreasing—Stann.

Gradually increasing, suddenly ceasing—Arg. met., ign., sul.

Hat, sensitive to pressure of—Glon.

Hair, on touching the—Agar.

Head, thinking fatigues the—Cocc.

Head, when moving—Caps., cimici., cocc. graph., ipec., lach., lob., nat. mur., phos., puls., sep., spig.

Head, must be constantly moved to and fro, and eyes closed—Agar.

Head, when shaking—Ign., sep.

Head feels weak, can hardly hold it up—China.

Head cannot be held upright, or raised, it is too heavy—Cocc., puls.

Head inclined to drop forward, could only be raised with difficulty, after which it would again drop forward—Nux m.

Head so heavy is obliged to hold it upright, in order to relieve the weight pressing forward in the forehead—Rhus.

Head involuntarily jerks backward and forward —Sepia.

Head involuntarily nods, when writing, as if some one pressed it down—Caust.

Head backward, inclined to bend—Cham., china, glon.

Head, unable to lie on back of—Cocc.

Head trembles—Cocc., ign.

Hemicrania, sees half light, half dark—Glon.

Hungry during headache—Psor.

Inspiration, during a deep—Anac.

Inspiration, during—Carbo v.

Intermittent—Ars., caul., cham., coloc., croc., plat.

Laughing, when—Phos.

Lie down, obliges one to—Croc., fer., mag. mur., phos. ac., rhus.

Lie down and be quiet all day, must—Gels.

Looking into the air, headache on—Thuja.

Looking down, headache on—Olean., spig.

Lying down, when—Bell., camph., coloc., euph., lyc.

Lying on the part affected, when—Graph, phos. ac.

Lying on the back, when—Coloc.

Mastication, from—Sulph.

Madness, drives to—China, cham.

Motion, during—Acon., agnus, am. mur., anac., aur., bry., cal. carb., carbo v., china, cocc., coloc., croc., dulc., kalm., lob., nat. mur., nux v., spig., stan., sulph., ther.

Motion, pain compels—Fer., phos. ac.

Move, on beginning to—Ther.

Music, from—Phos.

Noise, from—Anar., ign., merc., phos. ac., spig.

Paroxysms, in—Ant. crud., ars., bapt., caul., dig., ign., ipec., laur., plat., sang., verat.

Position, tries in vain to shift pain by change of—Ign.

Pregnancy, during—Cocc., nux v., psor., puls., sep.

Pressure, from—Agar., am. carb., phos. ac.

Pulse, pain with every—Glon.

Pulse, pain synchronous with—Camph., glon.

Raising the head—Bov.

Rising from sitting—Sulph.

Rising up—Acon., nux v.

Reading—Arg., arn., calc. carb., ign., nat. mur., opi.

Rest, during—Agar., am. carb., phos. ac., stan.

Room, in warm—Arn.

Room, in a—Arn., ars., coff., laur., zinc.

Room, on coming into—Colch.

Running, when—Nat. mur.

Rubbing, compels—Croc.

Seated, when—Agar.

Sitting up, compels—Cocc.

Sitting up, with head bent forward, compels—Ign.

Side, cannot lean or rest on left—Ars.

Side, when turning to it pain goes to other—Puls.

Scream, forces her to—Sep., sil.

School-girls, headache of—Nat. mur., phos. ac.

Shifts to the side on which he lies, pain—Phos. ac.

Sleep, after sound—Hepar.

Sleep, during—Cham.

Sneezing, from—Kalm.

Speaking, when—Acon., china, coff., dulc., ign., sil., spig.

Speaking, when hearing another—Ign.

Stand or walk, must—China.

Stand still when walking, must—Bell.

Stepping, when—China, nux v., phos., rhus, sep., sil., spig., sulphur.

Step, from a false—Anac.

Stairs, when going up—Ant. crud., ars., bell., cimici., ign., lob., paris, phos. ac.

Starting, causing—Crocus.

Stars, sees countless—Cyclam.

Stooping, when—Acon., ant. crud., bry., calc. carb., camph., coloc., dig., hepar, ign., lach., laur., nux v., paris, petr., plat., puls., rhus, sep., sil., spig., thuja, verat.

Sun, standing in—Glon., lach., nat. mur., nux v.

Sun, going into heat of—Glon., nat. carb.

Suddenly appears—Bell., gels., ign.

Suddenly ceases, after having gradually increased—Arg., ign.

Suddenly appears, slowly disappears—Sabina.

Suddenly appears, lasts indefinitely, ceases suddenly—Bell.

Touch, from—Bell., calc. carb., camph., caust., china, ipec.

Tread lightly, forces to—Chel.

Urine, passes off with a profusion of pale, limpid —Acon., ign., gels., verat.

Walking, when—Arn., caps., china, cocc., nux v., puls., sulph., ther.

Walking, quickly—Bell., bry.

Walking in the wind—China.

Wrapping up head, on—Calc. carb.

Writing, when—Calc. carb., nat. mur.

Yawning, when—Agar.

TIMES.

Three A.M.—Bovista.

Early in the morning—Hepar., nat. mur.

Awakens in the morning—Nat. mur.

Awakening in the morning, on—Croc.. graph., hepar, kali bi., kalm., nat. mur., nux v., phos. ac.

In the morning in bed—Agar., am. carb., anac., bry., calc. carb., caust., cham., graph., hepar, ign., lach., nat. mur., nux v., puls., zinc.

In the morning—Agar., am. carb., anac., arn., ars., aur., bov., bry., calc. carb., caust., cham., china, croc., graph., hep., kali bi., kalm., merc. iod. flav., nat. carb., nat. mur., nux v., petr., phos., puls., sep., sil., sulph., thuja., zinc.

In the morning, on rising—Am. mur., cyclam., kalm., lach., mag. mur., nux v., psor., puls.

In the morning after rising—Am. mur., hepar, puls.

In the morning after breakfast—Lyc., nux m.

After rising, lasting till toward noon, returning again at seven P.M.—Chel.

Begins in the morning, increases till noon, or a little later, then gradually decreases—Sulphur.

Morning till noon—Ipec., nat. mur., phos., sepia.

Morning till afternoon—China.

Every morning—Hepar, sepia.

Begins in morning, increases till noon, ceases toward evening—Kali bi., nat. mur., sang., spig.

Begins in the morning, increases through the day, growing milder toward evening—Nux v.

Appears at ten A.M.—Nat. mur.

Lasts till ten A.M.,—Nat. mur.

During the morning—Cimici., nat. mur.

At noon—Graphites.

Every day at noon—Arg.

Gradually increasing after dinner—Zinc.

After dinner till daylight next morning—Lob.

Afternoon, during—Bell., carbo v. , coloc., fer., graph., iris, lach., lyc., lob., sang., sil.

Afternoon, during part of—Cimici.

Beginning in afternoon, continuing into night—Verat.

Four to eight P.M.—Lyc.

Four P. M. till three A.M.—Bell.

Daily from four P.M. till morning—Bell.

Returning every evening—Dulc.

Evening—Am. carb., anac., bov., carbo v., cham., cocc., euph., graph., hepar, lach., puls., sulphur.

Towards evening in damp weather—Dulc.

Vanishes with the sun—Nat. mur.

Evening by candle-light—Crocus.

Lasts till evening—Lob.

Towards evening—Iris.

Evening after lying down—Puls.

Evening till midnight—Lob.

Evening till morning—Colch.

Evening in bed—Ars., lyc., puls., sep., sulph., zinc.

Comes and goes with the sun—Kali bi., nat. mur.

Increases and decreases with the sun—Kalm., nat. mur.

Increases and decreases with the sun every day —Glon., spig.

Repeated attacks during the day, or appearing at intervals of many days—Iris.

During the day—Am. mur., aran., cimici., hepar.

Lasts all day—Calc. carb., can. sat., chel., nat. mur., sep., stan.

Awakens at night—Hepar, nat. mur.

Awaking at night, when—Chel.

Midnight till morning—Hepar.

Midnight till ten a.m—Nat. mur.

Midnight, after—Phos. ac.

At night in bed—Ars., laur., merc.

At night—Am. carb., ant. tart., arn., ars., bov., cact., calc. carb., caust., cham., china, hepar, lyc., merc., phos. ac., phos., puls., rhus, sil., sulphur, zinc.

Persists uninterruptedly for two or three days— Crocus.

Begins with the warm weather and lasts all summer—Glon., nat. carb.

Constant—Can. sat., dulc., lob., nat. mur., phos., sepia.

Periodical—Aran., arn., ars., bell., calc. carb., cedron, china, cimici, fer., gels., ign., laur., lob., nat. mur., naja, nux v., rhus, sang., sil., spig., sulphur.

Every day—Ars., bell., calc. carb., lach., nat. mur., nux v., sep., sil., sulphur.

Every day at same hour—Ars., cimici., gels.

At regular hours—Aran.

Alternate days—Ars., china, phos.
Alternate days at certain hours—Nat. carb.
Every three or four days—Aurum.
Every seventh day—Sil., sulphur.
Every two weeks—Sulphur.
During full moon—Nat. carb.
With clock-like regularity—Cedron.

ACCESSORY SYMPTOMS.

Agitation—Ign., lyc.
Anguish—Cham., phos.
Appetite, loss of—Cocc., sel.
Asthma—Coloc., ipec.
Cheeks, redness of—Acon., cham., lach., nux v.
Coffee, desire for—Nux m.
Colic, with—Acon., cocc., chion.
Consciousness, with loss of—Nux v.
Constipation, with—Lac. deflor., pall., nux v.
Conversation, aversion to—Thuja.
Coryza with—Acon., lach.
Cranium, feeling as if the, were too small—Bell.
Cries, pain extorts—Coloc., sepia.
Deafness, with—Dulc.
Dejection, with—Ign., ther.
Delirium, with—Bell. nux v.
Discouragement, with—Agar., phos ac.
Distraction, with—Caps.
Ears, hammering in—Spig.
Ears, shooting in the—Merc., rhus.
Ears, humming in the—Acon., ars., aran., dulc.,
puls., sulph.
Epistaxis—Ant. crud., coff., dulc.
Eructations, with—Calc. carb., nat. carb., pall.,
nux v.

Extremities, as if beaten, pain in—Aconite.

Eyes, with affections of the—Crocus, opium.

Eyes, closing of the, with—Agar., bell., gels., nat. mur., oleand., sep., sulphur.

Eyes, with pains in the—Ant. tart., bry., cocc., croc., epiph., kali bi., lyc., nat. mur., puls., sil., stan.

Eyes, sore, bruised feeling in—Chion.

Eyes, with sparks before the—Lach.

Eyelids, with drawing of the—Bell.

Face, with heat in the—Calc. carb., can. sat., aran., lob., nux v.

Face, with pain in the—Sil.

Face, with paleness of the—Acon., phos., verat.

Face, with redness of the—Acon., cann., ign., plat., thuja, usnea.

Face, with yellowness of the—Lachesis.

Fainting, with—Graph., lyc., puls.

Fingers, with coldness of the—Canth.

Frantic, drives—Melilotus.

Heart, with palpitation of the—Ant. tart., hepar.

Head, pains compel the moving of the—China.

Indifference, with—Puls.

Irascibility—Bell., kali bi., sil., stan., thuja.

Jaws, with trembling of the—Carbo v.

Lie down, with desire to—Bell., bry., calc. carb., fer., lach., lyc., nat. mur., nux v., petr., phos. ac., rhus, sel., sil., sulphur.

Lying down, with inability to remain—Coloc.

Melancholy—Sel.

Mind, confusion of—Epiph.

Moans, with—Bell., ars.

Nape of the neck, with numbness of the—Spig.

Nape of the neck, with stiffness of the—Arg., graph., lach., spig., verat.

Nausea—Acon., am. carb., arg., ars., bry., calc. carb., camph., caps., carbo v., caust., chion., china, cocc., coloc., croc., epiph., graph., ign., ipec., iris., kali bi., kalm., lach., lob., nat. carb., nat. mur., nux v., phos., plat., puls., stan., sulph., thuja, verat., zinc.
Odontalgia, with—Rhus.
Perspiration in the head, with—Aconite.
Perspiration, with general—Natrum.
Perspiration, with cold—Graphites.
Photophobia, with—Euphr., kali bi., phell., puls.
Prostration, with general—Hydroc. acid.
Run hither and thither, with impulse to—Ars., coloc.
Shivering, with—Lach., nux v., sil., thuja.
Shuddering, with—Pulsatilla.
Sight, with confusion of the—Arg. nit., chion., cyclam., epiph., iris, gels., hydroc. acid, ign., natrum carb., natrum mur., puls., sil., sulphur.
Sleep, with inclinations to—Lach., gels.
Smell, acuteness of—Phos.
Stool, great desire for a—Chion.
Stomach, with pain in the—Verat.
Stomach, with pains in the pit of—Arg.
Vertigo, with—Æth., anac., ars., hydroc. ac., kali bi., kalm., lach., lyc., nux v., phos., puls., spig.
Vomiting, with—Acon., ant. tart., ars., bry., caps., chion., china, coloc., epiph., graph., ipec., iris, lach., lac deflor., lob., lyc., merc., nat. mur., nux v., phos., puls., plat., sep., suph., verat., zinc.
Weakness, with—China., lac deflor., lob., nux v., sil., sulph.
Weep, disposition to—Ars., plat.

SENSATIONS.

EYES :

As if the eyes would be torn out—Cocc.

As if the eyes were starting from their sockets—Bell.

As if receiving a blow on the eye—Agnus.

As if something lay above the eyes, and can not therefore look up—Carbo an.

As if something forcibly closed the eyes—Cocc.

As if a draft of air blew into the eyes—Cocc.

As if a thread were drawn tightly through the eyeballs and backward into the middle of the brain—Paris.

As if the head would swell up, and temples and eyes were pressed out—Paris

As if the eyes were pressed outward—Psor., puls., sabina, sang.

As if bruised, sore—Chion.

FOREHEAD :

Of a weight pressing through, or eyes—Aconite.

Pain in, as if the brain had not room enough—Psor.

As if a pulse was beating in—Kalmia.

As if would split—Olean., sang.

As if would be torn out—Hepar.

As if would burst—Ferr., nat. carb.

As if brain would press through, or eyes—Aconite.

As if brain were pressing against—Dulc.

As if brain would press through—Bell., bry., china.

As if forehead and vertex would be pressed asunder—Ranun.

As if forehead were compressed—Arn., can. sat., coloc., china.

As of a band compressing—Ant. tart., chel.

Constricted feeling of the skin of—Æscul., arn., bapt., caul., caust., paris, sabina.

Feeling of constriction in—Asclep., anac.

As if a rubber band was stretched over—Lilium.

Tension of scalp across forehead and temples—Caust.

Aching in, as if bruised—Hepar.

Aching in, like a boil—Hepar.

Headache, as from a heavy blow on the—Psorinum.

As of numbness in—Mag. mur.

Surging in, like waves of pain welling up and beating against frontal bone—Sepia.

As if a dull point were pressing in left frontal eminence—Crocus.

As if something was forced between frontal bone and cerebrum—Causticum.

As of an empty space between, and brain—Causticum.

Feeling of cold at small spots on the—Arnica.

TEMPLES :

As if a nail were thrust in—Arnica.

As if would burst—Lachesis.

As if were screwed together—Lyco.

Pain in both, as from strong pressure with thumbs—Cham.

As if the, would be crushed together—Caul.

As if the, were compressed—Cimici.

As if a sharp instrument were thrust through from one temple to the other—Asclep.

Ticking like a watch in the right—Chel.

Pinching in right, as if something pressed on the part and then relaxed—Opium.

SIDES OF THE HEAD :

Pressing from both sides, as if in a vise—nat. mur.

As if brain were moving to and fro towards the side which he stoops—Am. carb.

Undulation and whizzing, as of boiling water, on the side upon which one rests—Magn. mur.

As if a nail were thrust in side—Coff., ign., nux v.

As if a nail were thrust in left side—Agar., nat. mur.

As if a plug were driven into left side—Anac.

As if a convex button were being pressed on the part, on the left side—Thuja.

Brain feels loose on stooping, as if it fell to the left side—Nat. carb.

VERTEX :

As of something loose in the head, diagonally across top of it, when turning—Kalmia.

As if top of head would fly off—Bapt., cimici.

As of a ball driven from neck to vertex—Cimici.

Fullness in, as if the brain would burst out—Psorinum.

As if a nail were thrust in the—Thuja.

Pain on the, as if pressed asunder—Ranunculus.

As if the head were open along the—Spigelia.

As of a heavy weight on the—Cactus., can. sat., laur., nat. mur., phell., verat.

As of crackling in—Coffee.

As if the hair was pulled—Magn. mur.

OCCIPUT :

Seems fastened to pillow—Chel.

As if a weight were attached to occiput, pulling head backward—Agar.

As if the head were opening and shutting—Cocc.

At every step a jolt downward, as from a weight in the—Bell.

As of ice on the—Phosphorus.

As if a wedge were being pressed in the—Bovista.

As of a heavy blow on the—Can. ind.

As of electric sparks extending to the—Cham.

Pain, like a flash of lightning on the—Sang.

Furry sensation at the—Arnica.

BRAIN :

Violent shocks pass through the brain—Can. ind., glonine.

As if the brain was expanding—Glonoine

As if the brain were dashed to pieces—Coffee.

Brain feels too large—Arg. nit., cimici.

As if brain would press through forehead or eyes —Acon.

As if the brain would press through the forehead —Bell., bry.

As if the brain would press from within outward —Cimici.

As if the brain would be pressed asunder—Nux v.

As if the brain were crushed—Phos. ac.

As if the brain had not room enough—Psorinum.

Fullness in vertex as if the brain would burst out —Psorinum.

As if the brain were torn to pieces—Verat.

As if the brain was being torn—Ars., am. mur., coffee.

As if the brain were too full—Caps.

As if the brain were moving to and fro—China, crocus.

As if the brain beat against the skull—Ars., china, nux. v.

As if the brain beat against the skull in waves—China.

Waving sensation in the brain—Bell., china, cimici., glonoine.

As if on shaking the head the brain were loose and beat against the skull—Rhus.

Shaking in the brain, worse from motion—Spigelia.

Excessive sensitiveness of the brain to motion and walking—Nux v.

As if the brain were striking against the skull, on stooping—Laur.

As if the brain were vacillating—Lyc.

As if a ball were ascending—Acon., plumb.

As if the brain were being compressed into a ball—Arn. ant. tart.

As if the brain was in motion, when he leaned against something—Cyclamen.

As if the brain were moved by boiling water—Aconite.

As if the brain were loose—Lachesis.

As if the brain were loose and shaken, on walking—Nux v.

As if something were alive in the brain —Petr. sil.

Sensation at every step as if the brain rose and fell—Bell.

As if the brain would split—Am. mur., puls.

As if the brain were pressed together at both sides and out at the forehead—China.

As if something hard pressed on the brain—Ignatia.

Brain feels contracted—Laur.

As of an empty space between the forehead and brain—Caust.

As of a load on the brain—Arsen.

As if the brain were rolled up in a lump—Arnica.

As if thousands of needles were pricking into brain—Tarant.

As if a thread were tightly drawn through eyeballs and backward into middle of brain—Paris.

As of a ball rising from the throat into the brain—Plumbum.

CEREBELLUM :

As if laced together in cerebellum—Camphor.

SKULL :

As if the skull would split open—Bell., caps., nux v., spig.

As if the occiput was broken off from the rest of —Chel.

As if skull would burst—China, euphr., hepar.

As if bones of head were being driven asunder—Lyc.

As if calvarium was being lifted—Can. indica.

As if cranium was opening and shutting—Can. ind., cimici., cocc.

As if skull were cracking—Merc. iod. flav.

Cranium swells up in hard lumps—Sil.

As if cranium were too small—Bell.

Contractive pain, as from a band around cranium, with sensation as if the flesh were loose—Sulphur.

As if the skull were bruised, penetrating through

all the bones down to the root of the tongue—
Ipecac.

As if the brain were sticking against the skull, on
stooping—Laur.

As if the brain were loose and struck against the
skull, on moving head—Rhus tox.

As if the brain beat against the skull—Ars., china,
nux m.

As if the brain beat against the skull in waves—
China.

Bones of the head feel as if scraped—Phos. ac.

SCALP :

Feels constricted—Nat. mur.

As if scalp were drawn too tight—Carbo v.

Tension of scalp, across forehead and temples—
Caust.

Sensitive even to pressure of hat—Sil.

HAIR :

Feels as if standing on end—Acon., lob.

As if the hair were roughly grasped by the hand
—China.

As if the hair were being pulled on top of the
head—Nat. car., mag. mur.

NECK :

As of a heavy blow on back of the head and
neck—Can. ind.

As if the nape were broken—Graphites.

HEAD :

As if the head would drop to pieces if shaken—
Glonoine.

As if the head would split—Merc.

As if the head would burst—Bry., cact., calc. carb., caps., kali bi., lach., mag. mur., nat. mur., sang., sep., sulph.

Head feels as if it could be lifted off—Ther.

Swashing in the head as of water—Bell., hepar.

Vibration in the head when stepping hard—Sil.

As if a large quantity of cold water were poured on the head—Tarant.

Undulations through the whole head—Indigo.

Throbbing, like little hammers striking—Nat. mur., psorinum.

As if drops of water were falling on the head—Can. sat.

Head feels too large—Gels.

Head feels as if it were enlarged—Dulc., nat. mur., ranun.

Head feels as if it were being distended from within outward—Arnica.

Head feels full and as if expanding—Nux m.

Head feels as if it were getting larger—Merc.

Head feels distended—Bov., cedron.

Feels immensely large—Glonoine.

Feels swollen—Baptisia.

Feels as if pounded full of something—Cimicifuga.

Feels thick—Theridion.

Sensation of tightness around—Oleandra.

Sensation of a band around head above the ears—Gels, æth., merc., sulph., ther.

Sensation as of a bandage around the head—Carbo v., cocc., cylam., iris, merc., spigelia.

Sensation as if head was screwed together—Cocc., platina.

Sensation as if head was screwed up—Am. mur.

Sensation as if head were compressed in a vice—Cactus, merc., puls.

Feels too light—Gelsemium.

Feels too small—Coffee.

Pressure as from a stone in the head—Chamomilla.

Hat presses on the head like a heavy weight, feels the sensation even after taking it off—Carbo veg.

Feeling as if a board was on head—Æsculus hip.

Feeling as if a board pressed against the forehead—Dulcamara.

Feeling as of a heavy weight on the head—Arn., Phos. ac.

Feeling as if something heavy were sinking down into the head—Nux v.

Feeling as if a heavy weight pressed the head down on pillow—Merc. iod. flavus.

Feels heavy, as if it had something else upon it—Theridion.

As if a knife were drawn through the head—Arnica.

As if head was touched by sharp ice—Agaricus.

As if head were pierced by cold needles—Agaricus.

As if an abscess were forming in the head—Hepar.

As if a current of air were rushing through the head—Aurum, puls.

As if warm air was streaming up spine into the head—Arsenicum.

As of heat at small spots on the head—Arnica.

As if head was hollow—Argentum.

Feeling of emptiness in the head—Arg. met., cocc., nat. mur.

Muddled sensation in the head—Carbo v., lilium.

Dullness of the head, as if enveloped in a fog—
Petroleum.
As if a heavy black cloud enveloped the head—
Cimicifuga.
Stupid sensation in head—Asar., asclep.
Numb sensation in head and face—Bapt., graph.

GENERAL SENSATIONS :

Pain comes in terrific shocks—Sepia.
Sensation as if intoxicated—Arg. nit., bell.,
camph., caust., china, cocc., dulc., nux m., puls.,
rhus.
Sensation as if in a thick, smoky atmosphere—
Agnus.
Sensation as if floating in air—Nux mos.

LOCATION.

SUPRA-ORBITAL :

Pain in general—Am. carb., bell., cedron, cimic.,
colch., crocus, gels., lach., lil., magn. mur., merc.,
nat. mur., naja, sang., spig., stan., ther.
Aching—Laur., naja.
Beating—Lachesis.
Boring—Cimici., colch., laur.
Crampy—Colch.
Drawing—Can. ind., ign.
Dull—Can. ind., carbo v., kali bi., lil., rhus.
Heavy—Carbo v., kali bi., naja.
Pressive—Arn., arg. nit., aur., carbo v., china,
puls., sep.
Severe—Ars., cimici.
Soreness—Agnus.
Stitching—Anac., caps.
Tearing—Agnus.

Tensive—Glonoine.
Throbbing—Carbo v., glon., kali bi., naja.

LEFT SIDE :

Pain in general—Ign. phos.
Pressive—Caul., nux m.
Pulsating—Spig.
Shooting—Cedron.
Stinging—Cham., sel.
Stitching—Calc. carb.
Tearing—Cham.
Throbbing—Carbo v., glon., kali bi., naja.
Violent—Acon.

RIGHT SIDE :

Pain in general—Kalm.
Aching—Sil.
Boring—Sep.
Hammering—Sep.
Pressive—Arn., ign., sil.
Stitching—Anac., sep.

FOREHEAD :

Pain in general—Ant. crud., ant. tart., anac., bapt., bell., bry., camph., caps., caust., chel., colch., croc., gels.,ign., kali bi., kalm., lil., mag. mur., naja, nux v., olean., petr., psor., sabin., sang., selen., sil., stan., zinc.
Aching—Ant. crud., asar., arn., can. ind., caps., cocc., euph., hepar, kalm., lach., laur., lob.
Beating—Am. carb., aur., sil.
Boring—Ant. crud., arg. nit., aur., bell., cimici., cyclam., dulc., hepar, spig.
Bruised—Ars., puls., ranun.
Burning—Aur., bry., china, dulc., lil., phos.

Burrowing—Hepar.

Bursting—Gels.

Coldness—Arn., bell., cimici., laur.

Constriction—Paris., plat.

Crampy—Plat.

Cutting—Arg. nit., bell.

Darting—Æsculus hip., chel., cyclam.

Digging—Ant. tart., sang.

Drawing—Ant. tart., can. ind., carbo v., ign.

Dullness—Æscul., agar., ant. crud., can. ind., cocc., eup., iris, lil., lob., merc. iod. flav., phos., sabin., hyd. ac.

Fullness—Acon., cimici., laur.

Heat—Aran., carbo v., cimici.

Heaviness—Acon., am carb., am. mur., calc. carb., hepar, hyd. ac., iris, phos.

Jerking—Bry., can. ind., stan.

Lacerating—Bov., ipec.

Lancinating—Sang.

Numbness—Mag. mur.

Oppressive—Agar., calc. carb., lob.

Paroxysmal—Ant. crud., croc.

Piercing—Acon., sang.

Pressing—Acon., æscul., asar., arg. nit., arn., alu., am. carb., am. mur., agn., aur., anac., ant. tart., bell., camph., can. sat., carbo v., cham., chel., china, coloc., cyclam., dig., dulc., graph., ign., lach., merc., nat. mur., olean, paris, phos. ac., plat., psor., puls., ranun., rhus, sab., selen., sep., zinc.

Pricking—Aur.

Pulsating—Asar., am. carb., carbo v., kalm., mag. carb., phos.

Severe—Chion., glon., hyd., merc., petr., sep.

Sharp—Juglans.

Shooting—Bell., iris, kali bi., natr. mur.

Soreness—Ars., juglans.

Splitting—Ant. crud.

Sticking—Arn., bry., coloc.

Stinging—Lilium.

Stitching—Arn., aur., chel., cyclam., lach., sep., sulph.

Stupefying—Acon., ant. crud., ant. tart., bov., calc. carb., hepar, dulc., phos. ac.

Tearing—Am. mur., agn., anac., ant. crud., asar., arn., bell., carbo v., cham., chel., lyc., nat. carb., nat. mur., sulph., thuja.

Tensive—Ant. tart., caust., clem., glon., hepar, laur., nat. carb., nux v., sabin., sil.

Throbbing—Acon., alu., arg. nit., ars., bry., can. ind., cedron, chel., dig., iris, lach., merc. iod. flav., nat. mur., lac deflor., melil., phos., sang., sil., ther.

Violent—Acon., ant. crud., ars., asclep., bell., carbo v., cimici., cyclam., kali bi., plat., plumb., ther.

LEFT SIDE :

Pain in general—Asar., euph., nat. carb.

Boring—Arg. nit.

Dullness—Croc.

Pressing—Arg. nit., croc., iod.

Severe—Chion.

Stitching—Lach.

Thrusting—Croc.

RIGHT SIDE :

Pain in general—Iris.

Compressive—China.

Drawing—Ars.

Dullness—Cham., ign., ranun.

Pressing—Ars., caust., cham., chel., colch., nat. sul., opi.

Stitching—Cham.
Tearing—Calc. carb.
Throbbing—Ant. tart.

TEMPLES :

Pain in general—Caps., cham., chel., epiph., gels.,
lach., lil., lyc., nat. carb, naja., nux m., puls., sabin.,
stan.
   Aching—Asar., can. ind., caps., lyc.
   Beating—Lach.
   Boring—Arg. nit., cyclam., dulc., lac deflor.,
phell., sep., thuja.
   Burning—Arg.
   Coldness—Agar.
   Constrictive—Paris.
   Compressive—Asar., cimici.
   Crampy—Zinc.
   Cutting—Arg. nit., hyd.
   Darting—Chel., cyclam.
   Drawing—Agar., hepar, nux v., ranun.
   Dull—Æscul. hip.
   Flying—Æscul. hip.
   Fullness—Acon., cimici. lil.
   Griping—Mag. mur.
   Hammering—Sepia.
   Heat—Cimici.
   Jerking—China.
   Piercing—Acon.
   Pressive—Acon., arg. nit., agn., anac., arn., aur.,
can. sat., cham., coloc., cyclam., dig., glon., hepar,
lach., lob., opi., ranun, samb., sulph.
   Pulsating—Cact., chel.
   Severe—Caul., lach., sang., sep.
   Sharp—Bapt. hyd., merc. iod. flav.
   Shooting—Arn., bell., can. ind.

Soreness—Sang.
Stabbing—Bell.
Sticking—Arn., cham., paris.
Stitching—Æscul. hip., caust., china., cyclam.,
lyc., mag. mur., merc. iod. flav., sep., sulph.
Stupefying—Ant. tart.
Tearing—Agn., am. mur., anac., cham., chel., mag.
mur., nat. carb., puls., sulph., thuja., zinc.
Tensive—Ant. tart., caust.
Throbbing—Acon., arg. nit., caps., cedron., chel.,
cimici., glon., merc. iod. flav., nat. carb., phos.,
stan.
Tingling—Sulphur.
Violent—Cimici., cyclam.
Wandering—Cham.

LEFT :

Pain in general—Cyclam., lob., psor.
Burning—Coloc., merc.
Dullness—Lilium.
Drawing—Ant. crud.
Jerking—Stan.
Pressing—Ant. crud. aur., coloc., lil.
Pulsating—Ant. crud., spig.
Sticking—Calc. carb.
Stitching—Am. mur., arn.,
Tearing—Cham., sep.
Throbbing—Coloc.

RIGHT :

Pain in general—Chel., kali bi., sang.
Boring—Coloc., hepar.
Confusion—Aran.
Crawling—Plat.
Dullness—Cham,

Drawing—Agar., ant. crud., ant. tart.
Pressing—Can. ind., cedron, cham., chel., paris.
Pulsating—Spig.
Shooting—Kali bi.
Stitching—Can. ind.
Throbbing—Hepar.
Thrusting—Croc.

SIDES :

Pain in general—Kali bi.
Bruised—China, bov.
Compressive—Mag. mur.
Cutting—Arn.
Drawing—Dig.
Pressive—China, nat. mur.

SEMILATERAL :

Pain in general—Acon., agar., ars., arn., bell., calc.
carb., coff., coloc., cyclam., glon., graph., ign., ipec.,
kali bi., lach., nux v., plat., phos., puls.
Aching—Caps., kali bi.
Beating—Ars., euphr., sil.
Boring—Iris.
Bruised—Nux v.
Darting—Kali bi.
Dull—Iris.
Hammering—Iris.
Numbness—Plat.
Pressive—Caps., nux v., plat.
Pulsating—Crocus.
Shooting—Bell., can. ind., iris.
Stitches—Am. mur., nux v., puls.
Tearing—Aur., bell., merc., puls., sulph.
Throbbing—Cham., iris.

LEFT SIDE :

Pain in general—Am. mur., bov., calc. carb., lach., lob., sang., sep., spig., thuja.
   Coldness—Asar., lob.
   Confusion—Coloc.
   Cutting—Arg. nit.
   Digging—Arg. nit.
   Lacerating—Colc.
   Pressing—Agar., camph.
   Pulsating—Croc.
   Severe—Ars.
   Shooting—Chel.
   Stinging—Cham., sep.
   Stitching—Anac., ant. tart.
   Tearing—Agar, arn., caps., cham., sep.

RIGHT SIDE :

Pain in general—Bell., chel., lach., merc. iod. flav., nat. mur., sep.
   Beating—Lach.
   Boring—Bell., coloc.
   Bruised—Nux v.
   Burning—Coloc.
   Digging—Sang.
   Dragging—Arg. nit.
   Dull—Cedron.
   Heaviness—Chel.
   Lancinating—Sang.
   Piercing—Sang.
   Pressing—Agar., arn., kalm.
   Severe—Cact., coloc.
   Sharp—Mag. mur.
   Shooting—Bell.
   Sticking—Mag. mur.
   Tearing—Bell., lyc., magn. mur.

Throbbing—Sang.
Violent—Merc. iod. flav.

LEFT SIDE :

Pain in general—Alu., am. mur., bov., calc. carb.,
lach., sang., sep., spig., thuja.
  Coldness—Asar., lob.
  Confusion—Coloc.
  Cutting—Arg. nit.
  Digging—Arg. nit.
  Pressing—Agar., camph.
  Pulsating—Croc.
  Severe—Ars., arg.
  Shooting—Chel.
  Stinging—Cham., sep.
  Stitching—Anac., ant. tart.
  Tearing—Agar., arn., caps., cham., sep.

VERTEX :

Pain in general—Alu., agn., am. carb., bry.,
calc. carb., caust., coff., kali bi., lil., mag. mur.,
merc. iod. flav., pall., sil.
  Aching—Calc. carb., carbo v., cimici., lach., sil.
  Beating—Am. carb., lach., sil.
  Boring—Arg. nit., cyclam., spig.
  Burning—Arn., lach., sulph.
  Coldness—Laur., nat mur., sil., verat.
  Compressive—Graph.,
  Contractive—Sepia.
  Crampy, China.
  Cutting—Arg. nit.
  Darting—Cyclam.
  Digging—Sang.
  Drawing—Carbo v.
  Dull—Cact., cimici., hyd., lach.

Fullness—Cimici., psor.
Heat—Aur., carbo v.
Heavy—Cact., chel , ind.
Lacerating—Bov.
Lancinating—Dig., sang.
Oppressive—Graph., laur.
Pressive—Acon., am. carb, anac., arg. nit., cact.,
cedron, cham., croc., cyclam., carbo v., sulph., hyd.,
lyc., ranun., sep., sil., sulph., thuja, zinc.
Pulsating—Am. carb., nat. carb., sil.
Severe—Arn., bell., caps., merc., phos. ac.
Sharp—Merc. iod. flav.
Shooting—Can. ind.
Soreness—Zinc.
Stitching—Am. mur., caps., caust., cyclam.
Stupefying—Bov.
Tearing—Anac., aur., carbo v., lach. laur., nat.
carb.
Throbbing—Acon., anac., bry., caust., cimici.,
glon., nat. carb., sang.
Tingling—Sulph.
Violent—Cyclam., carbo v., dig.
Weight—Cact., nat. mur., phell.

OCCIPUT.

Pain in general—Am. mur., ars., bry., calc. carb.,
chel., china, cocc., gels., ipec., lach., naja, petr.,
pic. ac., rhus, sang., sil., zinc.
Aching—Graph., iod., sil.
Beating—Alu., ind.
Boring—Bell.
Burning—Graph., phos.
Cutting—Bell.
Drawing—Naja.
Dull—Æscul. hip., carbo v., cimici., gels., lil., petr.

Hammering—Camph.
Heat—Æscul. hip.
Heaviness—Chel., nat. mur.
Lancinating—Dig.
Pressive—Anac., arn., bell., bov., chel., lil., lob., petr., sil.
Pulsating—Hepar, petr., phos.
Severe—Glon., merc., nux m., nux v.
Sharp—Cedron.
Shooting—Bell., can. ind., nat. mur.
Shooting—Cimici., ipec.
Stinging—Ind., phos.
Stitching—Æscul. hip., chel., hepar, nat. mur.
Stunning—Can., ind.
Stupefying—Dulc.
Tearing—Am. mur., anac., bell., nux v., thuja.
Tensive—Chel., ipec.
Throbbing—Camph., glon., ign., ther.
Violent—Dig.
Weight—Nat. mur.

BRAIN :

Pain in general—Arg. nit., ars., bapt., bov., camph., cimici., cocc., iris, merc. iod. flav., opi., phos., spig.
Boring—Spig.
Bruised—Verat.
Burning—Arn., verat.
Coldness—Bell., phos.
Contractive—Laur.
Contusive—Nux v.
Digging—Dulc.
Dull—Merc. iod. flav.
Jerking—Stan.
Numbness—Plat.

Oppressive—Opi.
Pressive—Agar., ars., asar., bry., can. ind., china, ign., opi., verat.
Pulsating—Sepia.
Shocks—Can. ind., glon.
Soreness—Bapt., gels.
Stitching—Alu., bry.
Stupefying—Stan.
Tensive—Opium.
Tearing—Agar., ars., coloc., opi.
Throbbing—Bell., calc. carb., cham., psor.

BONES OF HEAD :

Aur., caps., hepar, ipec, kali bi., lyc., merc., phos. ac., sil., sulph.

PERIOSTEUM :

Phos. ac.

SCALP :

Aching—Hepar, hyd.
Biting—Olean.
Bruised—Ipec., petr.
Burning—Ars., bry., coloc., hepar, merc., olean., thuja, sulph.
Coldness—Agar., agn., calc. carb., chel., laur., sulph. verat.
Crawling—Arn., can. sat., chel., rhus.
Drawing—China, mag. mur., petr., phos. ac., puls., rhus, sep., thuja.
Eruption, painful—Graph., hepar.
Heat—Bell., bry., calc. carb., cham., coloc., verat.
Sensitive—Am. carb., ars., asar., ant. tart., bov., bry., calc. carb., caps., carbo v., china, coloc., graph., hepar, ign., kalm., lach., merc., nat. mur.,

nux v., paris., petr., rhus, sang., spig., sil., thuja, verat.

SCALP :

Soreness—Alu., carbo v., gels., olean., zinc.
Smarting—Coloc.
Tearing—Bry., carbo v., graph., lyc., rhus, sep.
Tension—Agn., arn., asar., caust., lach., merc., phos., spig.
Ulceration, as from subcutaneous—Ars., petr., phos. ac., rhus, zinc.

HAIR :

Sensitive to touch—Am. carb., bell., china, sep., sulph.
Sensitive—Alu., asar., calc. carb., caps., china, paris, sulph., thuja, verat.
Sensitiveness of roots of hair—China, sep., sulph.
Sensitiveness, as the hair was being pulled—Acon., alu., china, ind., rhus, sel.

WHOLE HEAD :

Pain in general—Arg. nit., caust., china, coloc., epiph., gels., lach., lil., sel.
Aching—Laur.
Beating—Ind.
Bruised—China.
Burning—Psor.
Coldness—Camph.
Confusion—Arn.
Dull—Epiph.
Dullness—Bapt., lil., epiph.
Heat—Nat. mur.
Pressive—Arn., croc., dig., puls.
Soreness—Glon.

Stitching—Merc.
Stupefying—Laur.
Throbbing—Glon., carbo v.

DIRECTION.

EYES :

Through head—Cimici.
To forehead—Cimici.

SUPRA-ORBITAL.

To base of brain—Cimici.
To crown—Arg. nit.
To eyes—Cimici.
To whole forehead—Stan.
To temples—Arn.

LEFT :

To whole forehead—Stan.
To occiput—Naja.

FOREHEAD :

Backward—Arn., bry., kali bi., lil., ther.
Deep into brain--Croc.
To eyes—Asar., phos.
Through, from right—Æscul. hip.
To back of head, and downward—Lil.
Over whole head—Anac., selen.
To nose--Sepia.
To root of nose—Phos.
To neck—Lyc.
To orbits—Chel.
To occiput--Bry., bell., cham., nat. mur., ther.
To temples—Cimici.
To vertex—Cimici.

RIGHT SIDE :

To parietal bone—Bell.

LEFT SIDE :

To eye—Ant. tart.
To occiput—Nat. carb.

TEMPLES :

Deep into brain—Croc.
To forehead—Cedron.
To temples, around forehead, just above eye-
brows—Lob.
To upper jaw—China.
To occiput—Cham.
To sides—Am. mur.
Temple to temple—Bell., chel., china. puls.
Flying pain through temple—Cham., æscul. hip.
Right to left temple—Lilium.
Right to zygoma—Ant. tart.
Left temple to occiput—Lilium.
Left temple to side—Sepia.
Both temples to root of nose—Agar.
Both sides to temples—Can. ind.
Both sides to vertex—Can. ind.

ONE SIDE:

To neck and shoulders—Lach.
To teeth and neck—Merc.
Though temples from one side to the other—
Alu., china., phos., sang.

RIGHT SIDE:

To eye—Mag. mur.
To right side of neck—Chel.
To left side—Arn.

LEFT SIDE:

To eye—Croc.; To eyes, zygoma and teeth—Spig.
Forward—Ant. tart.
To right side—Arn.
To occiput—Chel.
To forehead—Caps., caust., cham.
Upward—Glon.

OCCIPUT :

Down back—Graph.
To chest—Graph.
To eyes—Gels., glon., petr., sil.
Spreads upwards and settles over the right eye—Sang.
To ears—Chel.
To forehead—Arg. nit., caps., chel., nat. mur., petr.
Over whole head—China.
To lower jaw—Cham.
To neck—Graph.
Down back of neck—Cimici.
To sinciput—Calc. carb.
To shoulders—Bry., gels. ipec.
Down spine—Cimici., lil., nat. mur.
To temples—Glon.
Running in rays upward—Sang.
To vertex—Cimici., calc. carb.

NECK ; NAPE OF :

To eyeballs—Gels.
To forehead—Gels.
Over whole head—Carbo v., gels.
To occiput—Chel., dulc.
To shoulders—Ipec.

To supra-orbital region—Carbo v., sil.
To temples—Kalm.
Running in rays upward—Sang.
Upward and forward—Calc. carb., caust., cimici.,
gels., sil.
To vertex—Cimici., kalm., sil.

GENERAL :

Extends from above downward—Chel., ipec.,
phos. ac.
Base of brain to occiput—Cimici.
Below upward—Glon.
Behind forward—Phos. ac.
Behind forward in brain—Bell., bry.
Behind forward through eyeballs—Spig.
Brain to occiput—Can. sat.
Ear to parietal bone—Ign.
Ear to occipital protuberance—Ign.
Eyeballs backward—Lil., lach., phos.
Head to tips of fingers—Camph.
Head into left eye—Ign.
Left to right—Nux m.
Behind mastoid process to occipital protuberance
—Ign.
Right to left, over bridge of nose, left half of
face and forehead—Euphr.
Margin of orbits to temples—Can. sat.
Orbits to occiput—Cedron.
Right to left, completely around head—lach.
Skull to root of tongue—ipec.
Within outward—Acon., alu., arn., cimici., nat.
carb.
Within outward in forehead—Acon., alu., bell.,
bry., chel., olean., spig., sulph,

Within outward in temples—Cham., glon., lil., lob., lyc., sulph.

Within outward in side—Sepia.

Without inward—Ant. tart., cocc., plat., psor.

Without inward, in forehead—Am. carb., lach., ranun.

Without inward, in temples—Hepar.

Without inward, in vertex—Am. carb.

Extends through brain to occiput—Can. sat.

Whole brain—Coloc.

Head to supra-orbital region—Carbo v.

Head to nape of neck or chest—Nat. mur.

Extends to ears, root of nose, malar bones and jaws—Rhus.

Eyes, nose and teeth—Lyc.

Right side of head—Chel.

Whole head—Cimici.

CHARACTER OF PAIN.

ACHING—Arn., ipec., iris, laur., lyc., phos. ac., spig,
In forehead—Ant. crud., asar., arn., can. ind., cocc., corn., dulc., euphr., hepar, kalm., lach., laur.

In temples—Ant. crud., arn., asar., can. ind., lyc.

In vertex—Ant. crud., calc. carb., carbo v., cimici., hepar, lach., sil.

In occiput—Cimici., graph., iod., lob., rhus, sil.

In brain—Stan.

BEATING—Acon., ars., carbo v., euphr., glon., ind., ipec., lach., laur., nat. mur., puls., rhus.

In forehead—Am. carb., aur., lach., laur., mag. mur., sil.

In temples—Lach.

In side—Ars., lach.

In vertex—Am. carb., lach., laur., sil.
In occiput—Alu., ind.

BORING—Ant. crud., arn., bell., cyclam., coloc., laur., merc., nux v., sabin., spig.
  Supra-orbital—Cimici., laur., spig.
  In forehead—Ant. crud., arg. nit., aur., bell., cyclam., dulc., hepar, spig.
  In temples—Ant. crud., arg. nit., coloc., cyclam., dulc., hepar, sep., thuja.
  In side—Bell., coloc, iris.
  In vertex—Ant. crud., arg. nit., china, cyclam., spig.
  In occiput—Bell.

BRUISED—Arn., aur., bov., caps., china, coff., euphr., graph.
  In forehead—Ars., hepar, puls.
  In side—China, nux v.
  In vertex—Caust., mag. carb.
  In occiput—China.
  In brain—Carbo v., nux v., verat.
  In skull—Caps., ipec.

BURNING—Acon., agar., arn., ars., kali bi., lil., merc., nux v., phos., psor., rhus, thuja, verat.
  In forehead—Aur., bry., china, dulc., lil., phos.
  In temples—Coloc., merc.
  In sides—Coloc.
  In vertex—Arn., calc. carb., graph., lach., sulph.
  In brain—Arn., ars., verat.
  In scalp—Olean.

BURROWING—Hepar.
  In forehead—Hepar.

BURSTING—Caps., nat. mur.
  In forehead—Gels., nat. carb.

COLDNESS—Agn., bell., calc. carb., camph., kalm., laur.
   In forehead—Arn., bell., cimici.
   In side—Lob.
   In vertex—Agar., nat. mur., sil., verat.
   In occiput—Chel.

CRAMPY—Coloc., plat.
   In forehead—Plat.
   In temples—Zinc.
   In vertex—China.

CRAWLING—On temple—Plat.
   On scalp—Arn., can. sat., chel., coloc.

COMPRESSIVE—Asar., coloc., mag. mur., merc. iod. flav.
   In forehead—Arn., can. sat., chel., coloc.
   In temples—Asar., cimici.
   In sides—Mag. mur.
   In vertex—Graph.

CONTRACTIVE—Camph., laur., lyc., sulph.
   In vertex—Sepia.

CONSTRICTIVE—Graph., plat.
   In forehead—China, paris, plat.
   In temples—Paris.
   In occiput—China, graph.
   In scalp—Carbo v., china, nat. mur., plat., rhus.

CUTTING—Arn., bell.
   Supra-orbital—Hyd.
   In forehead—Arg. nit., caps., hyd.
   In temples—Arg. nit., bell., caps.

DARTING—Arn., caps., lach., mag. carb., spig.
   In forehead—Chel.

In temples—Arn., chel.
In sides—Kali bi., ranun.

DIGGING—Sabin., sang.
In forehead—Ant. tart., bry., sang.
In sides—Arg. nit.
In brain—Dulc., sang.

DRAGGING—Laur.
In side—Arg. nit.

DRAWING—Bapt., coloc., ipec., lil., plat.
Supra-orbital—Ign.
In forehead—Ant. tart., ars., can. ind., caps., ign., plat.
In temples—Agar., ant. tart., caps., hepar, nux v., ranun.
In side—Ant. crud., bell., dig.
In vertex—Carbo v.
In occiput—Arg. met., chel., graph., naja.

DULL—Ant. crud., ant. tart., asclep., bapt., bov., can. ind., caps., carbo v., caul., caust., cedron, cornus, chel., china, dig., dulc., gels., hepar, hyd., ipec., iris, kalm., lach., laur., lob., mag. mur., merc. iod. flav., nat. mur., naja, petr., phos., phos. ac., rhus., sabin., spig., sulph.
Supra-orbital—Æscul. hip., carbo v., cimici., gels., petr.
In forehead—Æscul. hip., agar., ant. crud., cham., cimici., lil., cocc., croc., euphr., hyd., ign., iris, lyc., merc., iod. flav., phos., puls.
In temples—Æscul. hip., ant. crud., cham., lil.
In side—Ars., cedron, iris.
In vertex—Ant. crud., anac., cact., cimici., hyd., lach.

In occiput—Æscul. hip., carbo v., cimici., gels. petr.

In brain—Merc. iod. flav.

FLYING—Cham., kali bi., puls.

FULLNESS—Æscul. hip., am. mur., arg. nit., bapt., bell., caps., cornus, lil., merc., nux m., psor., ranun.

In forehead—Acon., bell., cimici., laur., nat. carb.

In temples—Acon., cimici.

In vertex—Cimici.

HAMMERING—Am. carb., camph., ferr., hepar, nat. mur., psor.

Supra-orbital—Sepia.

In temples—Sepia.

In sides—Iris.

HEAT—Acon., ars., alu., camph., can. sat., caust., cocc., cyclam., glon., graph., ign., ipec., lach., laur., plat., rhus., sang., verat.

In forehead—Aran., carbo v., cimici., opi.

In temples—Cimici.

In vertex—Cimici., aur., carbo v., sulph.

HEAVINESS—Alu., ant. tart., arg. nit., arn., ars., bapt., bell., bry., camph., can. ind., carbo v., chel., china. cimici., croc., dulc., gels., ign., ind., ipec., iris., lach., laur., mag. mur., nat. mur., nux v., olean., opi., petr., rhus, stan., ther.

Supra-orbital—Carbo v., hyd., kali bi., lob., naja.

In forehead—Acon., am. carb., am. mur., calc. carb., hep., iris.

In temples—Acon.

In vertex—Cact., chel.

In side—Chel.
In occiput, Chel., merc. iod. flav., nat. mur.

JERKING—Merc. iod. flav., nat. mur., puls., sep.
In forehead—Bry., can. ind., stan.
In temples—China.
In occiput—Bell.
In cerebellum—Stan.

LACERATING—Alu., ind., ipec., lach., laur., puls., sulph.
In forehead—Ip., lyc.
In sides—Colch.
In occiput—Lyc.
In scalp—Colch.

LANCINATING—Am. carb., hepar, mag. carb., sang.
In forehead—Sang.
In vertex—Dig., sang.
In occiput—Dig.
In brain—Sang.

NUMBNESS—Acon., graph., mag. mur.
In forehead—Mag. mur.
In side—Plat.

OPPRESSIVE—In forehead—Agar., alu., calc. carb.
In vertex—Graph., laur.
In brain—Opi.

PRESSIVE—Æscul., ars., arn., aur., bapt., bell., bry., cact., caps., cham., china, cocc., coloc., cyclam., dig., gels., kalm., lach., nat. mur., nux m., nux. v., phos. ac., plat., puls., sabin., spig., stan.
Supra-orbital—Alu., arn., aur., carbo v., caul., china, ign., nux m., puls., sep., sil.

In forehead—Acon., agn., alu., am. carb., am. mur., anac., ant. tart., arn., ars., asar., bell., camph., can. sat., caps., caul., caust., carbo v., cham., chel., china., coloc. croc., cyclam., dig., dulc., graph., ign., iod., kali bi., lach., merc., nat. carb., nat. mur., olean., opi., paris, phos. ac., plat., psor., puls., ranun., sel., sep., stan., zinc.

In temples—Acon., agn., ant. crud., ant. tart., anac., arg. nit., aur., can. sat., caps., cedron, cham., chel., coloc., cyclam., dig., glon., hepar, lach., lil., lob., opi., paris, ranun., stan., sulph.

In vertex—Acon., arg. nit., am. carb., anac., cact., carbo v., cedron, cham., croc., cyclam., euphr., hyd., ranun., sep., sil., sulph., thuja, zinc.

In occiput—Arn., chel., petr., sil.

In brain—Agar., ars., asar., bry., china, ign., opi., verat.

PIERCING—Can. sat., paris, sang., sep.
In forehead—Acon., sang.
In temples—Acon.
In vertex—Acon., sang.
In brain—Sang.

PRICKING : In forehead—Aur.
In temples—Ant. crud.

PULSATING—Cact., glon., laur., nux m., opi., puls.
Supra-orbital—Spig.
In forehead—Am. carb., asar., carbo v., kalm., phos., sil.
In temples—Ant. crud., cact., spig.
In side—Croc.
In vertex—Am. carb., asar., carbo v., kalm., phos., sil.

In occiput—Hepar, petr., phos.
In cerebellum—Sepia.

SCREWING—Arg. nit.

SEVERE—Calc. carb., coloc., hyd., ind., kalm., sang.,
    ther.
  Supra-orbital—Ars., cimici.
  In forehead—Bell., glon., hyd., merc., sepia.
  In temples—Caul., lach., sang., sepia.
  In side—Ars., cact., coloc. ign.
  In vertex—Aur., bell., caps., merc.
  In occiput—Cimici., glon., merc., nux m.

SHOCKS—Can. ind., glon., lob., puls., sepia.

SEVERE—supra-orbital—Hyd.
  In forehead—Asar., iod.
  In temples—Bapt. hyd.
  In side—Bell., mag. mur.
  In vertex—Merc. iod. flav.
  In occiput—Cedron.
  In scalp—Cyclam.

SHOOTING—Acon., bell., cedron, kalm.
  Supra-orbital—Cedron, cornus.
  In forehead—Bell., iris, kali bi., nat. mur.
  In temples—Arn., bell., can. ind., chel., kali bi.,
    lil., merc. iod. flav.
  In side—Bell., can. ind., chel., iris.
  In vertex—Can. ind.
  In occiput—Bell., nat. mur.

SMARTING—Sabina.

SORENESS—Bry., glon.
  Supra-orbital—Agn.
  In forehead—Ars., bapt., paris.
  In temples—Sang.

In vertex—Zinc.
In occiput—Cimici., ipec.
In scalp—Alu., carbo v., gels., olean., spig.
In brain—Bapt., gels.

SPLITTING—Ant. crud.

STABBING—Sabin., spig.

STICKING—Paris.
In temples—Arn., cham., paris.
In forehead—Arn., bry., coloc.
In side—Caps., mag. mur.

STINGING—Acon., ipec., phos., sep.
Supra-orbital—Cham., sel.
In forehead—Lil.
In temples,—Cham.
In side—Cham., sep.
In occiput—Ind., phos.
In scalp—Cyclam., sep.

STINGING—Acon., ipec., phos., sep.
Supra-orbital—Cham., sep.
In scalp—Cyclam., sep.
In occiput—Ind., phos.

STITCHING—Arn., caps., carbo v., caust., cyclam.,
hepar., merc., nat. mur., nux v., paris, petr.,
puls., rhus, thuja.
Supra-orbital—Anac., ant. tart., caps., sep.
In forehead—Æscul. hip., am. mur., alu., arn.
aur., caps., chel., cham., cyclam., nat. carb.,
sep., sulph.
In temples—Æscul. hip., am. mur., calc. carb.,
can. ind., caust., cham., china, cyclam., lyc.,
mag. mur., merc. iod. flav., nat. carb., sep.,
sulph.

In side—Am. mur., anac., ant. tart., bry., nux v., puls.
In vertex—Am. mur., caps., caust., cyclam., nat. carb.
In occiput—Æscul. hip., chel., hepar, nat. mur.
In brain—Alu., bry.

STUPEFYING—Acon., ant. tart., arg. nit., bell., china, laur., mag. mur., nat. mur., nux v., phos., psor., puls., rhus, sep., stan., sulph.
In forehead—Ant. crud., ant. tart., bov., calc. carb., carbo v., dulc., hepar, nat. carb., phos. ac.
In temples—Alu., ant. tart.

TEARING—Am. mur., anac., ant. crud., ant. tart., ars., bell., caps., caust., cocc., coloc., croc., merc., puls., rhus, sep., thuja.
Supra-orbital—Agn., cham.
In forehead—Agn., anac., ant. crud., asar., aur., bell., calc. carb., caps., carbo v., chel., lyc., nat. carb., nat. mur., sulph., thuja.
In temples—Agn., am. mur., anac., ant. crud., cham., mag. mur., nat. carb., sep., sulph., thuja, zinc.
In side—Agar., arn., aur., bell., caps., lyc., mag. mur., puls., sep., sil.
In vertex—Ant. crud., aur., carbo v., lach., laur., nat. carb.
In occiput—Am. mur., anac., bell., nux v., thuja.
In brain—Agar., ars., coloc., opi.

TENSIVE—Ant. tart., lyc.
Supra-orbital—Glon.
In forehead—Ant. tart., caust., clem., glon., hepar, laur., nat. carb., nux v., sabi., sil.
In temples—Ant., caust.

In side—Plat.
In brain—Opi.
In occiput—Chel., ipec.

THROBBING—Anac., asar., bell., calc. carb., camp., can. ind., caps., carbo v., caust., china., cimici., cocc., cyclam., glon., iod., ipec., iris., kalm., lach., nat. mur., nux m., petr., puls., rhus, sang., sep., sulph.

Supra-orbital—Carbo v., glon., kali bi., lyc., naja, nux m., ther., usnea.

In forehead—Acon., alu., arg. nit., ars., bry., can. ind., caps., cedron, chel., cimici., dig., iris, merc. iod. flav., nat. mur., phos., sang., sil., ther.

In temples—Acon., arg. nit., caps., cedron, chel., cimici., coloc., glon., hepar, lach., merc. iod. flav., nat. carb., phos., stan.

In side—Bell., cham., iris.

In vertex—Acon., bry., caust., cimici., glon., nat. carb., sang.

In occiput—Chel., glon., ign., ther.

In brain—Bell., calc. carb., psor., sang.

TINGLING—Acon., sulph.

VIOLENT—Acon., ant. crud., ars., asar., asclep., cham., chel., china, cimici., cocc., coloc., cyclam., dig., euphr., gels., glon., graph., hepar, ind., iod., kali bi., kalm., lach., laur., lyc., merc. iod. flav., nat. mur., opi., phos., phos. ac., plat., sang., sep., sil., thuja.

In forehead—Ars., asclep., bell., cyclam., iris, kali bi., selen., sil., spig., theridon.

AGGRAVATIONS.

TIMES OF :

Three to four A.M.—Thuja.
Early in morning—Chel., croc.
In the morning, after rising—Am. mur., lil., mag. mur., sab.
In the morning—Ars., arn., aur., cham., cyclam., kalm., nat. carb., nat. mur., nux v., phos., sep., spig.
In the morning, on awaking—Caust., sulph.
In the morning, when washing—Anac., calc. carb.
Ten A.M.—Nat. mur.
Towards noon—Chel., sulph.
At noon—Can. ind., kali bi.
Noon till night—Caul.
Until two P.M.—Dulc.
After dinner—Chel., nat. carb.
Afternoon—Arn., chel., cyclam., ind., iod., thuja.
Afternoon till midnight—Lil.
Four P.M., till evening—Caust.
Four till eight P.M.--Lyc.
During day—Am. mur.
Towards evening—Lil., sepia.
Evening—Alu., agn., am. carb., ars., bry., cact., caps., caust., cham., chel., cyclam., ind., kalm., lach., laur., lil., nat. carb., nat. mur., puls., ranun., sulph.
Five to ten P.M.—Puls.
Evening by candle-light—Crocus.
Evening when undressing—Oleandra.
On going to bed—Alumen.
Evening in bed—Zinc.
Sitting up at night—Bovista.

During nightly fever—Hepar.
At night—Arg., ars., cact., lob., merc., thuja,

CONDITIONS OF :

Acids, from—Lach.
Air, from cold—Camph., cocc., dulc., iris., sil.
Air, from warm—Iodine.
Air, from open—Aur., cedron., china, coff., glon.,
    kalm., mag. mur., merc., nux v., spig., sulph.
Air, from walking in open—Alu., am. carb., arn.,
    bell., caps., ign., lil., stan.
Air, when going from, into room—Chel.
Air, from draught of—China.
Alcohol, from—Lach.
Awaking, on—Arn., croc., kalm., lach., merc.
    iod. flav., nat. mur.
Bending head forward, from—Coloc.
Bending head backward, from—Anac., glon.
Cold water, from applications of—Ars., calc.
    carb., glon.
Cold, from—Calc. carb., dulc.
Coffee, from—Cocc., ign., nux v.
Combing hair back, from—Rhus.
Conversation, from—Iodine.
Coughing—Anac., arn., bry., caps., ipec., nux v.,
    sulph.
Drinking, after—Cocc., merc.
Eating, when—Am. carb., ars., cocc.
Eating, after—Am. carb., anac., carbo v., china,
    coff., merc., nux m., phos., sulph., zinc.
Excitement, from—Paris.
Footsteps, jarring of room by—Sil., spig., ther.
Gaslight, from working under—Glon.
Grief, from—Ign., phos. ac.
Heat, from—Kalm., nat. mur.

Heat of sun—Calc. carb., glon., nat. carb.
Heat of bed—Bell., lyc., merc., rhus, sulph., thuja.
Hot and cold things—Merc.
Hair, from cutting—Bell., phos.
Head, from anything around—Gels
In-doors, when—Sepia.
Joy, from—Coffee.
Labor, from hard—Anac.
Laughing—Cocc.
Lemonade, from—Selen.
Light, from—Acon.,arg. nit., ars., bell., cact., caul.,
  cocc., coff., ign., kali bi., nat. mur., nux v.,
  sang., sil.
Looking upward, from—Caust.
Lying down, when—Aur., bell., china, gels., glon.,
  mag. mur., rhus.
Lying in bed—Merc.
Lying on back, when—Coloc.
Lying on painless side—Sepia.
Lying on painful side—Hepar.
Mental labor, from—Arg. nit., arn., calc. carb.,
  cham., china, dig., lyc., nat. carb., nat. mur.,
  nux v., petr., phos., sil., sulph.
Motion, from—Acon., agn., anac., ars., aur., bell.,
  bry., camph., chel., china, coloc., coff., cyclam.,
  dulc., epiph., glon., graph., hepar, iod., kali bi.,
  kalm., lach., lob., mag. mur., nux v., paris,
  ranun., sang., sep., sil., spig., sulph.
Motion in house—Sepia.
Motion at night—Nat. carb.
Motion, violent—Iris, phos.
Move, when beginning to—Iris, rhus.
Moving jaw, when—Phos.
Moving head,from—Arn., bell.,caps., cimici, croc.,
  glon., ipec., nat. mur., phos.

Moving eyes, from—Bell.,caps., chel., cimici., ign., nat. mur., nux v., opi., paris, sepia.
Moving upper eyelid—Coloc.
Music, from—Phos., phos. ac.
Noise, from—Acon., ars., bell., cact., caps., caul., cocc., coff., ign.. iod., kali bi., nux v., phos. ac., sang., spig., sil., ther.
Nose, when blowing the—Lilium.
Over-heating, from—Glon.
Odors, from strong—Aur., ign.
Pressure, from—Bovista.
Pressure, from, of hat—Glon.
Resting in a warm room, from—Plat.
Rest, from—Caps., cyclam., epiph., ipec., ind.., lil, puls., rhus, stan., sulph.
Reading,from—Aran., arn., caust., glon., nat. mur.
Riding—nux m.
Raising head, from—Stan.
Raising eyes, from—Ign., puls.
Rising from lying, on—Acon., ars., arn., ther.
Rising, on—Cedron, chel., kalm., lyc., merc. iod. flav., rhus.
Rising from sitting, on—Acon., æscul. hip., bell.
Room, in the—Coffee.
Room, in the warm—Alu., did., laur., merc. iod, flav., phos., zinc.
Shaking head—Asar., caust., cornus, glon.
Shock, from—Bell.
Sitting, from—Agar., caust., ind., puls.
Sitting up, when—Chel.
Sleep, from—Cocc., lach., merc.
Smoking—Cocc., lach., merc.
Sneezing, from—Sulph.
Standing, from—Coloc.

Stairs, on ascending—Alu., ant. crud , bell., calc. carb., glon., ign.

Stepping, from—Alu., bell., china.

Stooping, from—Æscul. hip., am. mur., arn., bry., caps., caul., caust., cham., coloc., cornus, cyclam., glon., hepar, ign., ipec., kali bi., lach., lyc., nux v., phos., plat., rhus, sil., spig., sulph., ther.

Stool, when straining at—Ign.

Suppression of eruption, from—Nux m., sulph.

Talking, from—Calc. carb., cact., cocc., dulc., nat. mur., sil.

Tobacco, from smell of—Ign.

Tea, from drinking—Selen., thuja.

Thinking—China, coff., ign., nux v., paris, phos., spig.

Thinking of the disease—Cham., ign., lilium.

Touch, from—Acon., aur., bell., china, kalm., merc., phos., rhus, sang., sil., sulph., thuja.

Turning, when—Kalm.

Turning head rapidly, when—Nat. carb.

Turning head quickly—Rhus.

Warmth, from—Arg. nit., arn., merc., nat. mur., petr.

Walking, from—Arn., asar., bell., calc. carb., caps., carbo v. china, cimici., clem., cornus, coloc., glon. nux v.

Walking fast, from—Chel., sep.

Walking, from getting warm while —Lyc.

Weather, from damp—Dulc., glon., rhus.

Weather, from damp, cold—Dulc, puls., sulph.

Weather, from wet, cold, stormy—Sulph.

Weather, change of—Dulc., nux m.

Weather, wet—Am. carb., dulc., nux m., puls., rhus.

Wind, from—Spig.
Wine, from—Glon., nux v., selen., zinc.
Writing, from—Aran., glon.

AMELIORATIONS.

TIMES OF :

Morning—Bovista.
Morning, on rising—Alu. nux v.
Noon, towards—Cedron.
During dinner—Anac.
Towards evening—Kali bi., nux v.
Evening—Anac., arn., spig.
At sunset—Lilium.
Night—Laur.

CONDITIONS OF :

Air, in open—Acon., ant. crud., aran., ars., aur.,
caust., cimici., coffee, glon., ipec., kali bi., laur.,
lob., lyc., nat. mur., phos., plat., puls., sabina,
selen.
Air, in cool open—Croc., iod., lyc.
Air, walking in open—Ant. crud., coloc., hepar,
sang., sep., thuja.
Bending head backward, from—Rhus.
Bending head forward, from—Gels., ign.
Binding something around head, from—-Arg. nit.,
calc. carb., puls.
Cold, from—Kalm.
Cold water, from drink of—Acon.
Cold applications to head, from—Calc. carb.
Cold water, from—Ars., calc. carb., caust., cy-
clam., iod., phos.
Contact, from—Coloc., cyclam.
Darkness, from—Sang., sep.

Eating, after—Bovista, chel., iod., psor.
Erect, when sitting erect after lying—Merc.
Eyes, from closing—Calc. carb., chel., sep.
Eyes, from opening—China.
Exercise, from gentle—Nat. mur.
Heat, from—Sil.
Looking at one point, from—Agnus.
Lying down, from—Bell., calc. carb., camph., china, dulc., epiph., glon., kali bi., ign., lach., lyc., nat. mur., nux v. olean, phos. ac.
Lying down at night, from—Arn., mag. mur.
Lying on painful side, from—Arn., ign., nux. v., sep.
Lying with head high—Caps., spig.
Lying down in a quiet dark place—Sil.
Lying on back—Ign.
Mental occupation, from—Merc. iod. flav.
Motion, from—Agar., aur., cham., cyclam., ind., phos. ac., plat., rhus, stan., sulph., thuja.
Motion, from moderate, continued, Iris.
Motion in open air, from—Asar., mag. mur., nux v.
Perspiration, from—Nat. mur., nux v., thuja.
Pressure, from—Agar., alu., am. carb., cact., carbo., chel., cimici., coloc., glon., hyd., ind., lil., merc. iod. flav., nat. mur., paris, puls., spig.
Pressure, from, hard—China, mag. mur., sang.
Pressure on forehead, from—Bell.
Pressing head against something hard—Kali bi.
Quiet, from—Cocc., sang.
Rest, from—Chel., epiph., sep.
Resting head on cushion, from—Alu.
Resting head on hand, from—Aran., lil.
Rising in morning—Alu., nux v.
Rising, on—Cham.

Rising, after—Hepar.
Room, in the—China, cyclam., merc., sulph.
Rubbing, from—Ars., ind., tar   huja.
Shaking head, from—Gels.
Smoking,—Aran.
Sitting, from—Ars., bell., coloc., cyclam., glon.,
  nat. mur., nux v.
Sitting with head and shoulders resting on high
  pillow, when—Gels.
Sleep, from—Epiph., glon., sang., sep., sil.
Sleeping at night, after—Ign.
Stool, after—Agar., cornus.
Stooping, from—Ign.
Vomiting, after—Asar., calc. carb., glon.
Washing, from—Psor.
Walking—Aur., caps.
Will, from effort of—Lilium.
Warmth, from—Ars., aur., caps., cocc.
Wrapping head up warm—Mag. mur., nux v.,
  sil., thuja.

INDEX.

213